Atomic and Molecular Wires

NATO ASI Series

Advanced Science Institutes Series

A Series presenting the results of activities sponsored by the NATO Science Committee, which aims at the dissemination of advanced scientific and technological knowledge, with a view to strengthening links between scientific communities.

The Series is published by an international board of publishers in conjunction with the NATO Scientific Affairs Division

A Life Sciences **B Physics**	Plenum Publishing Corporation London and New York
C Mathematical and Physical Sciences **D Behavioural and Social Sciences** **E Applied Sciences**	Kluwer Academic Publishers Dordrecht, Boston and London
F Computer and Systems Sciences **G Ecological Sciences** **H Cell Biology** **I Global Environmental Change**	Springer-Verlag Berlin, Heidelberg, New York, London, Paris and Tokyo

PARTNERSHIP SUB-SERIES

1. Disarmament Technologies	Kluwer Academic Publishers
2. Environment	Springer-Verlag / Kluwer Academic Publishers
3. High Technology	Kluwer Academic Publishers
4. Science and Technology Policy	Kluwer Academic Publishers
5. Computer Networking	Kluwer Academic Publishers

The Partnership Sub-Series incorporates activities undertaken in collaboration with NATO's Cooperation Partners, the countries of the CIS and Central and Eastern Europe, in Priority Areas of concern to those countries.

NATO-PCO-DATA BASE

The electronic index to the NATO ASI Series provides full bibliographical references (with keywords and/or abstracts) to more than 50000 contributions from international scientists published in all sections of the NATO ASI Series.
Access to the NATO-PCO-DATA BASE is possible in two ways:

– via online FILE 128 (NATO-PCO-DATA BASE) hosted by ESRIN, Via Galileo Galilei, I-00044 Frascati, Italy.

– via CD-ROM "NATO-PCO-DATA BASE" with user-friendly retrieval software in English, French and German (© WTV GmbH and DATAWARE Technologies Inc. 1989).

The CD-ROM can be ordered through any member of the Board of Publishers or through NATO-PCO, Overijse, Belgium.

Series E: Applied Sciences - Vol. 341

Atomic and Molecular Wires

edited by

Christian Joachim

CEMES-CNRS,
Toulouse, France

and

Siegmar Roth

Max Planck Institute,
Stuttgart, Germany

Springer Science+Business Media, B.V.

Proceedings of the NATO Advanced Research Workshop on
Atomic and Molecular Wires
Les Houches, France
6–10 May 1996

A C.I.P. Catalogue record for this book is available from the Library of Congress

ISBN 978-94-010-6485-9 ISBN 978-94-011-5882-4 (eBook)
DOI 10.1007/978-94-011-5882-4

Printed on acid-free paper

This book contains the proceedings of a NATO Advanced Research Workshop held within the programme of activities of the NATO Special Programme on *Nanoscale Science* as part of the activities of the NATO Science Committee.

Other books previously published as a result of the activities of the Special Programme are:

NASTASI, M., PARKING, D.M. and GLEITER, H. (eds.), *Mechanical Properties and Deformation Behavior of Materials Having Ultra-Fine Microstructures.* (E233) 1993 ISBN 0-7923-2195-2

VU THIEN BINH, GARCIA, N. and DRANSFELD, K. (eds.), *Nanosources and Manipulation of Atoms under High Fields and Temperatures: Applications.* (E235) 1993 ISBN 0-7923-2266-5

LEBURTON, J.-P., PASCUAL, J. and SOTOMAYOR TORRES, C. (eds.), *Phonons in Semiconductor Nanostructures.* (E236) 1993 ISBN 0-7923-2277-0

AVOURIS, P. (ed.), *Atomic and Nanometer-Scale Modification of Materials: Fundamentals and Applications.* (E239) 1993 ISBN 0-7923-2334-3

BLÖCHL, P. E., JOACHIM, C. and FISHER, A. J. (eds.), *Computations for the Nano-Scale.* (E240) 1993 ISBN 0-7923-2360-2

POHL, D. W. and COURJON, D. (eds.), *Near Field Optics.* (E242) 1993 ISBN 0-7923-2394-7

SALEMINK, H. W. M. and PASHLEY, M. D. (eds.), *Semiconductor Interfaces at the Sub-Nanometer Scale.* (E243) 1993 ISBN 0-7923-2397-1

BENSAHEL, D. C., CANHAM, L. T. and OSSICINI, S. (eds.), *Optical Properties of Low Dimensional Silicon Structures.* (E244) 1993 ISBN 0-7923-2446-3

HERNANDO, A. (ed.), *Nanomagnetism* (E247) 1993. ISBN 0-7923-2485-4

LOCKWOOD, D.J. and PINCZUK, A. (eds.), *Optical Phenomena in Semiconductor Structures of Reduced Dimensions* (E248) 1993. ISBN 0-7923-2512-5

GENTILI, M., GIOVANNELLA, C. and SELCI, S. (eds.), *Nanolithography: A Borderland Between STM, EB, IB, and X-Ray Lithographies* (E264) 1994. ISBN 0-7923-2794-2

GÜNTHERODT, H.-J., ANSELMETTI, D. and MEYER, E. (eds.), *Forces in Scanning Probe Methods* (E286) 1995. ISBN 0-7923-3406-X

GEWIRTH, A.A. and SIEGENTHALER, H. (eds.), *Nanoscale Probes of the Solid/Liquid Interface* (E288) 1995. ISBN 0-7923-3454-X

CERDEIRA, H.A., KRAMER, B. and SCHÖN, G. (eds.), *Quantum Dynamics of Submicron Structures* (E291) 1995. ISBN 0-7923-3469-8

WELLAND, M.E. and GIMZEWSKI, J.K. (eds.), *Ultimate Limits of Fabrication and Measurement* (E292) 1995. ISBN 0-7923-3504-X

EBERL, K., PETROFF, P.M. and DEMEESTER, P. (eds.), *Low Dimensional Structures Prepared by Epitaxial Growth or Regrowth on Patterned Substrates* (E298) 1995. ISBN 0-7923-3679-8

MARTI, O. and MÖLLER, R. (eds.), *Photons and Local Probes* (E300) 1995. ISBN 0-7923-3709-3

GUNTHER, L. and BARBARA, B. (eds.), *Quantum Tunneling of Magnetization - QTM '94* (E301) 1995. ISBN 0-7923-3775-1

PERSSON, B.N.J. and TOSATTI, E. (eds.), *Physics of Sliding Friction* (E311) 1996. ISBN 0-7923-3935-5

MARTIN, T.P. (ed.), *Large Clusters of Atoms and Molecules* (E313) 1996. ISBN 0-7923-3937-1

DUCLOY, M. and BLOCH, D. (eds.), *Quantum Optics of Confined Systems* (E314). 1996. ISBN 0-7923-3974-6

ANDREONI, W. (ed.), *The Chemical Physics of Fullereness 10 (and 5) Years Later. The Far-Reaching Impact of the Discovery of* C_{60} (E316). 1996. ISBN 0-7923-4000-0

NIETO-VESPERINAS, M. and GARCIA, N. (Eds.): *Optics at the Nanometer Scale: Imaging and Storing with Photonic Near Fields* (E319). 1996. ISBN 0-7923-4020-5

RARITY, J. and WEISBUCH, C. (Eds.): *Microcavities and Photonic Bandgaps: Physics and Applications* (E324). 1996. ISBN 0-7923-4170-8

LURYI, S., XU, J. and ZASLAVSKY, A. (Eds.): *Future Trends in Microelectronics: Reflections on the Road to Nanotechnology* (E323). 1996. ISBN 0-7923-4169-4

JAUHO, A. and BUZANEVA, E.V. (Eds.): *Frontiers in Nanoscale Science for Micron/Submicron Devices* (E328). 1996. ISBN 0-7923-4301-8

ROSEI, R. (Ed.): *Chemical, Structural and Electronic Analysis of Heterogeneous Surfaces on Nanometer Scale* (E333). 1997. ISBN 0-7923-4489-8

Table of Contents

Contacting the wire: Planar junction

Contacting the wire: STM junction

Transport theory

Preface

This volume contains the proceedings of the NATO Advanced Research Workshop on "Atomic and Molecular Wires". It was sponsored by the Ministry of Scientific Affairs Division special program on Nanoscale Science with the support of the CNRS and the Max Planck Institute. Scientists working or interested in the properties of wires at a subnanoscale were brought together in Les Houches (France) from 6 to 10 May 1996.

Subnanoscale wires can be fabricated either by surface physicists (atomic wires) or by synthetic chemists (molecular wires). Both communities present their foremost advances using, for example, STM to assemble atomic lines atom for atom, to fabricate a mask for such a line or using the wide range of chemical synthesis techniques to obtain long, rigid and conjugated oligomers.

Interconnecting such tiny wires to sources (voltage, current) continues to demand a great technological effort. But nanolithography associated with microfabrication or STM are now clearly identified paths for measuring the electrical resistance of an atomic or a molecular wire. The first measurements have been reported on Xe_2, benzene, C_{60}, di(phenylene-ethynylene) showing the need for a deeper understanding of transport phenomena through subnanowires. Such transport phenomena like tunnel (off-resonance) transport and Coulomb blockade have been discussed by theorists with an emphasis on the exponential decrease of the tunnel current with the wire length versus the ballistic regime of transport.

SELF ASSEMBLING OF METAL LINES ON METAL SURFACES

J.P. BUCHER
Institut de Physique et de Chimie des Matériaux de Strasbourg (IPCMS), UMR 46 du CNRS, Université Louis Pasteur
23 rue du Loess, F-67037 Strasbourg, France

ABSTRACT. The density, of kinetically grown islands on surfaces as well as their size and their shape rely on fundamental aspects of adatom diffusion. Highly elongated one-dimensional adatom islands can be grown spontaneously by controlling the anisotropic (temperature driven) diffusion of freshly deposited adatoms on a substrate with groves or channels, the simplest example of which is an (110)-fcc substrate. Beyond interesting applications in the field of nanostructures, an analysis by variable temperature STM provides fundamental information on relevant parameters like the migration barriers of adatoms. A few important conditions to be satisfied for the synthesis of metal lines on metal surfaces will be reviewed and some examples will be presented from the recent research. Finally, the possibility of getting self-organized islands on surfaces will be addressed and the driving mechanisms will be discussed.

1. Introduction

The fascinating possibility of growing atomic chains on surfaces has found much attention since the invention of STM, because the surface structure can be investigated in real space at an atomic scale, but also because artificial structures can be built by dragging atoms and molecules over the surface with the STM-tip [1,2]. Here we will be concerned with structures that form spontaneously on surfaces when the adsorbed metal vapor condenses into islands. The purpose of this contribution is to show under which conditions it is possible to create nanostructures with a surface density of up to 10^{14} per cm^2 with well defined dimensionalities and shapes. This question is particularly relevant to the synthesis of low dimensional systems, since the dimension of electronic components and sensors is tending more and more towards a limit where atomic and quantum phenomena play a leading role.

We will show that by controlling the kinetics it is possible to grow nanostructures with a well defined surface density and shape. Beyond the nice possibility of growing nanostructures, we will show that fundamental information on adatom migration barriers can be obtained from variable temperature STM experiments, with considerable implication for the growth of ultrathin films on metallic substrates. Then we will enumerate the necessary conditions to synthesize one dimensional structures on metals, we will stress the importance of phase separation in heterostructures and finally we will address the issue of self-organization in these structures and particularly, we will emphasize the role of strain relief.

C. Joachim and S. Roth (eds.), Atomic and Molecular Wires, 1-10.

2. Building nanostructures by controlling the growth kinetics

2.1. ISLAND DENSITY

Lets introduce some basics in the field of diffusion of atoms on a surface by means of the Ag on Pt(111) system which represents a very simple example a two-dimensional isotropic diffusion. Figure 1 shows variable temperature STM images of the structures grown and imaged at a given temperature. The evolution is shown for temperatures between 80K and 400K and the silver coverage is 0.1 ML. The continuous diagonal lines in the images are monatomic steps of the Pt(111) substrate and the bright zones are the monatomic thick silver structures. A remarkable observation is that the density of islands drops markedly when the temperature is increased while the average size of the islands increases. There is even a temperature (170K in this example) at which no islands are nucleated on the terraces anymore and silver condenses at the steps of the platinum substrate.

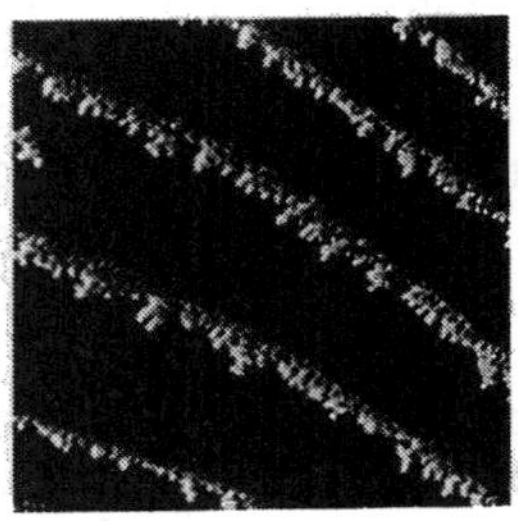

Fig. 1. STM topographs of low coverage Ag submonolayer structures (Θ = 0.1 ML) on a Pt(111) surface, grown and imaged at the temperature indicated. The lateral scale is given by the bar, representing a length of 500 Å [3].

When an incoming adatom is adsorbed on the surface from the gas phase it may have two distinct destinies. It may meet another adatom during its thermally activated random walk and form a stable nucleus (an immobile dimer), or it may be directly incorporated into an already existing island. The later operation does not contribute to increase the number of islands on the surface it just increases the size of an existing island. The first event is much more frequent in the first stage of the growth process while the incorporation into an already existing island will dominate at longer time. At very long time, no new islands are formed and an incoming adatom diffuses to an island before it has the opportunity to meet another adatom. As a result, the island density saturates and we can define a capture area for each island. The average separation between islands then provides a good measure of the diffusion length. Nucleation at steps occurs when the average diffusion length before encounter of an island becomes bigger than the extension of one terrace. An analytical expression of this saturation density of islands N_{sat} has been proposed by Venables [4]

$$N_{sat} = \eta(\Theta)\,(R/\nu_0)^{i/(1+2)} \exp[(E_i + iE_d)/kT(i+2)], \qquad (1)$$

where η is a slow varying function of the coverage Θ, R is the deposition rate and ν_0 is the attempt frequency. The size of the critical nucleus is i and E_i is its binding energy. When the critical nucleus is 1, E_i becomes zero and we get the exponent $E_d/3$. The formula reflects the fact that the adatom diffusion is thermally activated with an energy barrier E_d.

2.2. ADATOM MIGRATION BARRIERS

Migration barriers of atoms on surfaces can be obtained by exploiting the results of variable temperature STM. This has been done in Fig. 2 where the saturation density of islands as a function of 1/T has been reported in an Arrhenius plot, the straight segments correspond to temperature ranges with i=1 and i=2 respectively. The example of Fig. 2 nicely shows that in order to determine the diffusion barrier of a single adatom, it is of advantage to work at low temperature, in the regime where the critical nucleus is one, because then we don't have to worry about the binding energy of dimers, trimers etc... which are generally not well known. The slope of the low temperature segment directly provides the activation energy E_d. For Ag on Pt(111), the Arrhenius plot of Fig. 2 leads to a value of the adatom diffusion barrier of $E_d = 157 \pm 10$ meV. The attempt frequency ν_0 (prefactor of Eq. (1)), is comprised between 1×10^{13} Hz and 4×10^{14} Hz in good agreement with values from other experiments. At higher temperatures, above 120 K for this system, we have to consider that the critical nucleus is no longer i=1 but i=2.

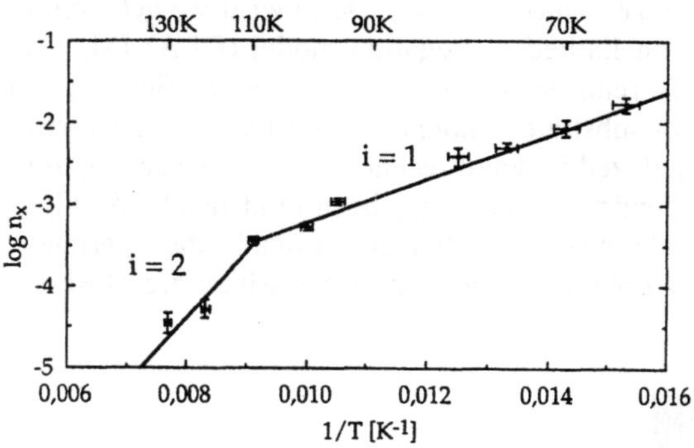

Fig. 2 Arrhenius plot of the saturation island density obtained with a deposition rate of $R = 1.1 \times 10^{-3}$ ML/s. Up to T = 110 K the critical nucleaus is i=1 and the straight line fitting the data allows to deduce the diffusion barrier E_d [5].

The existence of a low temperature critical nucleus has been ascertained from the following experiment [6]. When we deposit silver vapor on a Pt(111) surface at a temperature of 40 K, dimers will form spontaneously. Upon increasing the temperature of the substrate the dimers remain stable only up to 100 K then the islands grow either by dissociation of the dimers and diffusion of individual atoms or by diffusion and aggregation of the dimers themselves. This naturally introduces the notion of critical nucleus, (i+1) is the smallest island that is still stable at a given temperature.

We analyzed a situation in which growth was dictated by kinetics rather than by thermodynamics. Growth can also be dictated by nucleation at defects, for example Co on Au(111) nucleates at well defined sites, the elbows of the chevron reconstruction of Au(111). As a result, at higher Co coverage, compact islands align along rows defined by these defects [7].

2.3. ISLAND SHAPE

Up to now we have shown that the density of islands relies on the diffusion of adatoms on the bare surface. We show now that the shape of these islands relies on the diffusion of adatoms along the edges of the islands. At low temperatures, diffusing atoms stick where they hit, therefore islands will show fractal structures because diffusion is hindered along the perimeter of the island (Fig. 3).

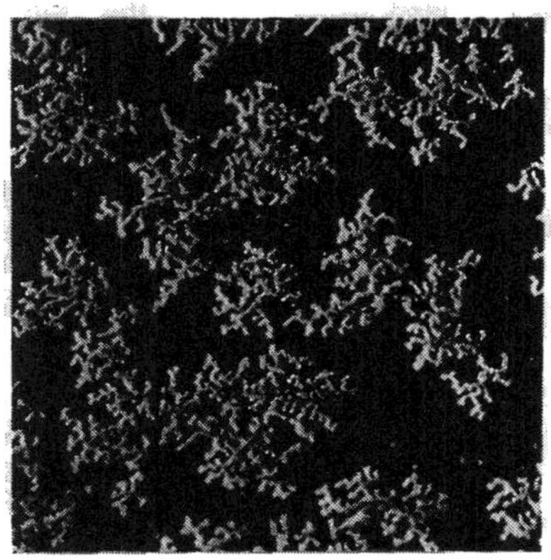

Fig. 3 STM image showing fractal aggregates grown on Pt(111) at 110 K and a deposition flux $R = 1.6\times10^{-5}$ ML/s (image 1200 Å × 1200 Å) [6].

The fractal dimension of the islands is 1.78 which is very close to what we get from the computer simulation within the diffusion limited aggregation model (DLA) [8]. This ideal situation of frozen perimeter, is realized in practice only at sufficiently low substrate temperatures. The incidence of substrate temperature and flux of the incoming adatoms, on the final shape has been analyzed in detail recently for this system [9] and a transition from randomly ramified to dendritic islands has been evidenced. At higher temperatures where diffusion of adatoms is activated around the perimeter, rearrangements take place and we may obtain compact structures which are close to a thermodynamic equilibrium.

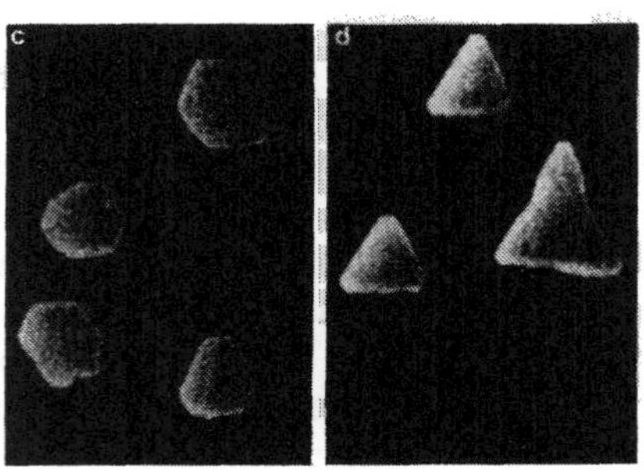

Fig. 4 Island shapes on Pt(111) after deposition at substrate temperatures of 455 K, and 640 K. STM images 770 Å x 1100 Å and 2300 Å x 3300 Å respectively and $\Theta = 0.15$ [10].

For example hexagonal aggregates have been obtained by depositing Pt atoms on Pt(111) at 450 K and it was shown by Michely et al. [10] (Fig. 4) that there is a transition from hexagonaly shaped islands to triangularly shaped islands as a function of growth temperature. They explained their results on the basis of the well known crystallographic fact that a compact island on a (111)-type of substrate is limited by both, edges with a {111} facet and edges with a {100} facet, the triangular shapes are explained by temperature dependent anisotropic diffusion along these two edges.

3. How to synthesize 1D structures on metals

The question that comes up next is the following: would it be possible to control the adatom substrate system in order to tailor the islands dimensionality? or could we make a 1D system out of a 2D system? The answer is yes, it was tried in different ways. A first attempt was to grow stripes on vicinal surfaces. Vicinal surfaces are surfaces that are obtained by a small miscut away from high density plane. This leads to a surface with an adjustable step density, all steps oriented along the same direction, in which it is possible to grow adatom stripes. The result was rather disappointing and the reason can be understood from Fig. 1d: the silver stripes at the Pt(111) steps are far from being perfect.

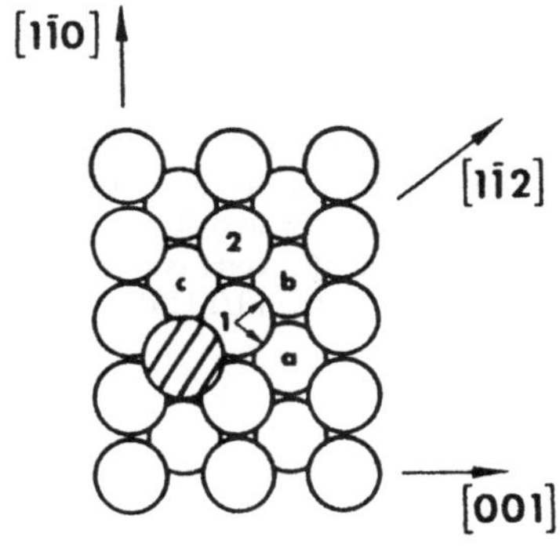

Fig. 5 Top view onto a (100)-substrate. Exchange displacement directions for an adatom (shaded) are shown as discussed in the text.

Here, we describe a technique that allows to grow atomic scale stripes without the need of steps. By choosing a crystal plane with a particular symmetry, we determine at the same time the topography of the surface. The details of the surface corrugation play an important role for the easy diffusion of adatoms along potential wells or grooves. We tried to exploit the anisotropic diffusion to tailor metastable islands of particular shapes. The simplest example of an anisotropic substrate is a (110) surface of a fcc-crystal (Fig.5). Such a substrate is made of compact atomic rows separated by channels along the [1$\bar{1}$0] direction. When an adatom lands on such a substrate, it will diffuse preferentially along the channel in a linear type of random walk, at least at not too high temperatures. We expect that linear chains will form spontaneously by aggregation of adatoms diffusing along the channels. Equation (1) is the result for a two dimensional system which applies well to surfaces that have no anisotropies. The previous formula, however, can be generalized for any dimensionality [11,12]:

$$N \sim (\Theta R/D)^{d/2(d+1)}. \qquad (2)$$

Here D is the temperature dependent diffusion factor, and d is the dimensionality. By introducing the value d = 2, we will get the usual exponent 1/3. In the case of linear diffusion (d = 1), the exponent is 1/4.

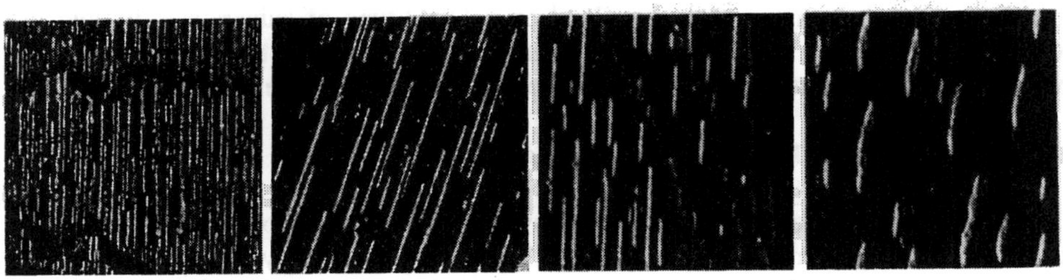

Fig. 6 STM images of Cu islands grown and imaged on Pd(110) at four different substrate temperatures and nearly the same Cu coverage $\Theta = 0.1$ ML. Temperatures are (a) 265 K, (b) 300 K, (c) 320 K, and (d) 350 K. The images 1200 Å × 1200 Å [13].

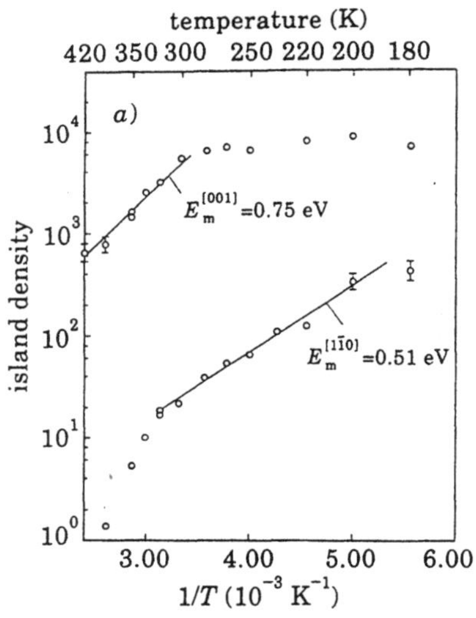

Fig. 7 Arrhenius plot of the 1D transverse density of islands (upper curve, $N=10^0$ corresponds to 420 islands/cm) and of the overall density of islands (lower curve, $N=10^0$ corresponds to 1.7×10^{10} islands/cm^2 as a function 1/T [13]

We have tested these expressions on a carefully chosen system, namely Cu on Pd(110). Figure 6 shows variable temperature STM images of Cu islands (0.1 ML) on Pd(110) that have been grown and imaged at temperatures ranging from 265 K to 350 K. At temperatures below 300 K, linear chains several 100 Å long, "monatomic" in width, grow spontaneously along the $[1\bar{1}0]$ direction. Above 300 K, 2D islands that are still elongated along the $[1\bar{1}0]$ direction start to form. There is a 1D to 2D transition while the surface density of islands drops markedly as a function of T. It was shown from LEED experiments that these Cu islands grow pseudomorphically on Pd(110). A better insight into what happens is obtained again by drawing an Arrhenius plot of the density of islands. Figure 7 shows the island densities in the $[1\bar{1}0]$ direction and in the [001] direction as a function of temperature as obtained from island counting on several

STM images for a wide range of temperatures (170 K to 420 K). The slope of the island density along the [1$\bar{1}$0] direction gives the diffusion barrier for atoms in that direction, it has a value of 0.51 eV. Interestingly the island density in the transverse direction remains constant for temperatures below 300 K, illustrating the fact that there is no transverse diffusion (across the channels) at these temperatures. The transverse diffusion is activated only above 300 K with a barrier of 0.75 eV.

From our energy barriers, we find that the jump rate of atoms in the easy direction is 20'000 times faster than across the channel walls, this explains to a large extent the anisotropic shape of the islands. However at temperatures above 300 K the diffusion is also activated along the perimeter of the islands as explained previously which may also alter their shape.

4. Phase separation

State of the art embedded atom model calculations show that the energy barrier for the atoms to hop across the wall would be of the order of 1.0 eV, which is much bigger than the experimental value of 0.75 eV. Therefore, an interesting question arises on the transverse migration mechanism and one may ask about the occurrence of a cross exchange [14]. In such a mechanism, the shaded Cu atom would exchange with a Pd atom of the compact wall (atom No 1 in Fig. 5) which will go to position either a) or b). The activation energy for such an exchange mechanism in the [1$\bar{1}$2]-direction is 0.57 eV, as found from molecular dynamics calculation [13]. Such a mechanism seems quite plausible and would explain the relatively low migration barrier observed experimentally. Assuming that the Cu atom would in turn be kicked out by a diffusing Pd atom this would provides a mechanism of Cu atom propagation with a lower activation energy. This interpretation, however, is hardly compatible with the phase separation of Cu and Pd observed from CO adsorption IR spectroscopy and also from photoemission experiments for temperatures below 700 K [15]. As a consequence, Cu atoms most probably occupy the top layer and therefore no alloying takes place.

Alloying is nevertheless a phenomenon we have to take care about if we like to synthesize well defined low dimensional structures on metal surfaces. Surface alloys have been observed even in systems that are known to be immiscible, there are various examples now in the literature [16,17]. Burrowing of Au atoms on Ag(100) surfaces [18] has been observed recently, in this case the Au atom digs a hole into the substrate. These are effects we certainly want to avoid in order to build well defined nanostructures on surfaces.

5. Self organized structures: the role of strain relief

Figure 8 shows an STM image with a Cu-coverage of 0.35 ML deposited at 300 K on Pd(110). While the monatomic Cu chains at very low coverage appear as randomly distributed, two-dimensional Cu chains at intermediate submonolayer coverage appear to be arranged in at least a semi-ordered striped structure. A part of the islands show even a completely regular arrangement (lower left part of Fig. 8).

To analyze the apparent mesoscopic order on a quantitative level, we have drawn histograms of the island distances along the [001] direction (Fig. 9). For a given Cu island, the distance to the first, second, third, and all other neighboring islands in the

[001] direction were measured for more that 300 line scans taken from 10 STM images. These histograms can be understood as one-dimensional island-island correlation functions.

The distribution for monatomic copper chains (Fig. 9a) shows indeed the absence of any long-range order, since all island-island distances are present with nearly equal probability. Only at a distance below 20 Å a drop of the occurrence probability is noticed, which might be interpreted as a repulsive island-island interaction, giving rise to a depleted zone of some Angstroms around the copper chains. The distance distribution of the two-dimensional islands at a coverage of about 0.35 ML (Fig. 9b), on the other hand clearly reveals the presence of some long-range order. Distinct maxima at island separations of 40 Å, 80 Å, 120 Å are present. These values are multiples of the average island separation which is about 40 Å, showing an enhanced probability to find neighboring islands at this particular position. The loss of order with increasing distance resembles the typical behavior for glasses or liquids.

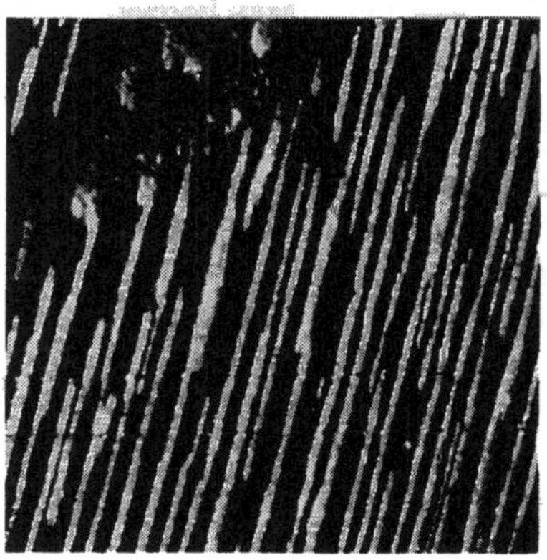

Fig. 8 STM image 1200 Å × 1200 Å showing the growth of Cu on Pd(110) at T = 300K and Θ = 0.35 ML [15].

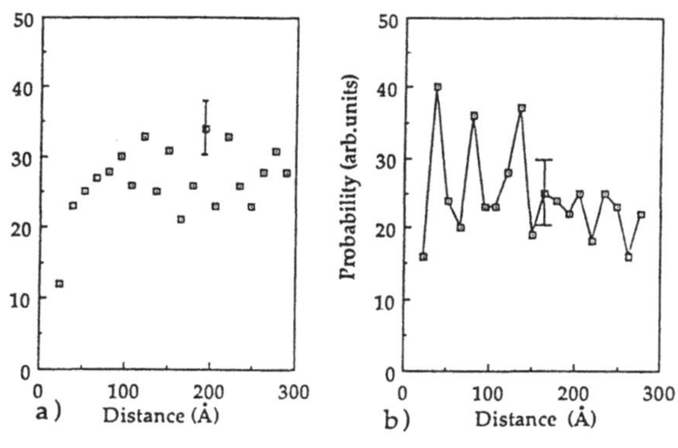

Fig. 9. Island-island separation histograms along the [001] direction upon Cu deposition at 300 K for (a) Θ = 0.07 ML, note the scattering of the data which is due to the statistics, (b) Θ = 0.35 ML [15].

Mesoscopic domain ordering was observed on Si(100) [19], on Au(111) [20] and in Cu(110) covered with oxygen [21]. The physical origin of these ordering phenomena are believed to be long range elastic and/or electrostatic interactions. As a matter of fact we must consider that the copper islands are under considerable tensile stress due to the 7 % lattice mismatch between Cu an Pd. The stressed Cu islands relax at the boundary and exert a force on the Pd(110) substrate which is elastically distorted.

The spontaneous formation of periodic domain structure was first studied by Marchenko [22] and then by others [23-26]. This approach has been used to explain the results of strained Si(100) surfaces [23], it relies on the fact that the strain relief is the driving force to promote the formation of long range ordered domain patterns. Let us consider alternating stripes of two phases A and B of width respectively L_A and L_B. The extra free energy per unit length ΔF due to the formation of domains is given by

$$\Delta F = 2F_S/(L_A + L_B) - [2C_{el}/(L_A + L_B)] \ln \{(L_A + L_B)(\sin \pi\Theta)/2\pi a\}, \quad (3)$$

$$\Theta = L_A/(L_A + L_B) \qquad 0 < \Theta < 1, \quad (4)$$

ΔF is the sum of two terms, the first term is the free energy per unit length for the creation of a boundary, the second term describes the elastic relaxation. The logarithm becomes infinity for the limits $\Theta = 0$ or 1 and it goes through a minimum for $\Theta = 0.5$. C_{el} depends on elastic parameters like the shear modulus and the Poisson ratio of the substrate as well as on the difference in the normal components of the surface stress between A and B domains and "a" is the lattice constant. The equilibrium periodicity D is obtained by minimizing ΔF with respect to L_A, keeping Θ constant

$$D(\Theta) = \kappa/\sin \pi\Theta = L_A(\Theta) + L_B(\Theta), \quad (5)$$

$$L_A(\Theta) = \kappa\Theta/\sin \pi\Theta, \quad (6)$$

with $\kappa = 2\pi a \exp(1 + F_S/C_{el})$. L_A does not vary much as a function of Θ for $0 < \Theta < 0.7$, and the D curve as a function of Θ is also rather flat around a central value of about half coverage. It has been shown that there should exist a firm relation between the periodicity D at an intermediate coverage Θ and the width of a single island L_A at low coverage [26]. The ratio D/L_A is comprised between 0.33 and 0.25 for a coverage of about 0.5. This ratio is roughly independent of C_{el} and the free energy for the formation of one boundary. In our case, island ordering is observed only above a critical coverage, somewhere between 0.15 and 0.25 ML. At a coverage of 0.2 ML, the average island width is 11 Å and the repeat distance is 40 Å, leading to a ratio of 0.28 which is within the predicted interval. This interpretation of island ordering is strictly valid only for systems at equilibrium. Although at a substrate temperature of 300 K, the shape of the copper islands is determined to large extent by the growth kinetics as was shown before, we cannot exclude kinetic pathways, which might also lead to semi-ordered structures during growth [27].

6. Conclusion

We have shown that by controlling adatom diffusion on surfaces, we can tune the surface density and shape of the adislands. The shape and the dimensionality of Cu islands on Pd(110) can be adjusted to a large extent by controlling the anisotropic diffusion of freshly deposited Cu atoms on the substrate. Cu islands grow preferentially along the $[1\bar{1}0]$ direction since diffusion of single adatoms occurs mostly along this fast diffusion direction. We observed a transition from one- to two-dimensional islands upon temperature increase during growth. These result show that the anisotropy of diffusion is at the very origin of the elongated islands. Beyond the nice possibility of building nanostructures, the variable temperature STM experiments provide fundamental information about adatom diffusion barriers. We have shown that phase mixing has to be avoided in order to grow well defined nanostructures on metal surfaces and that the dimensionality of islands can be adjusted on some carefully chosen systems by exploiting the anisotropic diffusion. Finally, we have shown that self organized structures may be obtained under certain conditions of strain relaxation. It must be stressed that the approach of diffusion-controlled aggregation is also applicable to other systems in the kinetic regime (including metals on semiconductors or insulators). The determination of the correct temperature window in which one- or two-dimensional aggregates can be grown is then the key issue.

Acknowledgements

I would like to thank K. Kern, E. Hahn, C. Massobrio and A. Pimpinelli for stimulating discussions.

References

1. D.M. Eigler and E.K. Schweitzer, Nature, **344**, 524 (1990).
2. M.F. Crommie, C.P. Lutz, D.M. Eigler, Science **262**, 218 (1993).
3. H. Röder, H. Brune, J.P. Bucher and K. Kern, Surf. Sci. **298**, 121 (1993).
4. J. Venables, G.D.T. Spiller and M. Hanbücken, Rep. Prog. Phys. **47**, 399 (1984).
5. H. Röder, Thesis, EPFL, Lausanne 1996.
6. H. Röder, E. Hahn. H. Brune, J.P. Bucher, K. Kern, Nature **366**, 141 (1993).
7. B. Voigtlander, G. Meyer and N.M. Amer, Phys. Rev. B **44**, 10354 (1991).
8. T.A. Witten and L.M. Sander, Phys. Rev. B **27**, 5686 (1983).
9. H. Brune et al. Surf. Sci. **349**, L115 (1996).
10. T. Michely, M. Hohage, M. Bott and G. Comsa, Phys. Rev. Lett. **70**, 3943 (1993).
11. Y.W. Mo, J. Kleiner, M.B. Webb, M. Lagally, Phys. Rev. Lett. **66**, 1998 (1991).
12. A. Pimpinelli, J. Villain, D.E. Wolf, Phys. Rev. Lett. **69**, 985 (1992).
13. J.P. Bucher, E. Hahn, P. Fernandez, C. Massobrio and K. Kern, Europhys. Lett. **27**, 473 (1994).
14. P.J. Feibelman, Phys. Rev. Lett. **64**, 729 (1990).
15. E. Hahn, E. Kampshoff, A. Fricke, J.P. Bucher and K. Kern, Surf. Sci. **319**, 277 (1994).
16. L.P. Nilsen et al. Phys. Rev. Lett. **71**, 754 (1993).
17. M.B. Hugenschmidt, A. Hitzke, R.J. Behm, Phys. Rev. Lett. **76**, 2535 (1996).
18. E.S. Hirschorn et al. Surf. Sci. **323**, L299 (1995).
19. F.K. Men, W.E. Packard and M.B. Webb Phys. Rev. Lett. **61**, 2469 (1988).
20. J.V. Barth et al. Phys. Rev. B **42**, 9307 (1990).
21. K. Kern et al. Phys. Rev. Lett. **67**, 855 (1991).
22. V.I. Marchenko, JETP Lett. **33**, 381 (1981).
23. O.L. Alerhand, D. Vanderbilt, R.D. Meade, and J.D. Joannopoulos, Phys. Rev. Lett. **61**, 1973 (1988).
24. D. Vanderbilt, in *Computations for the Nano-scale*, Eds. P.E. Blöchl et al. (Kluwer,New York 1993) p.1.
25. J. Tersoff and R.M. Tromp, Phys. Rev. Lett. **70**, 2782 (1993).
26. P. Zeppenfeld et al. Phys. Rev. Lett. **72**, 2737, (1994).
27. A. Wheeler, C. Ratsch, A. Morales, H. Cox, and A. Zangwill, Phys. Rev. B **46** 2428 (1992).

SELF-ASSEMBLED SINGLE ATOM WIDE METAL LINES ON Si(001) SURFACES

J. NOGAMI
Department of Physics and Laboratory for Surface Studies, Univ. Wisconsin-Milwaukee, P.O. Box 413, Milwaukee, WI 53201, USA

Group III metals form straight lines of adatom dimers on the Si(001) surface when deposited at room temperature. This has been seen in STM images of Al, Ga, In and Sn (group IV) on Si(001). Theory and experimental data from other techniques have converged on a structural model for these lines of metal atoms known as the parallel adatom dimer model. The current understanding of the characteristics of these lines is reviewed. The differences between a group III adsorbate (In) and a group IV adsorbate (Sn) are discussed. New STM results of the coadsorption of In and Sn are shown and discussed in light of the growth mechanism of these lines.

1. Introduction

When one considers the smallest possible lateral dimensions of a wire, the natural limit for a minimum cross sectional area is a single atom. This wire could be pictured as a single row of atoms, and ideally would be conducting, be supported on an insulating substrate, and have contacts at both ends so that its transport properties could be easily studied. Diverse materials and systems have been discussed at this conference that to varying degrees approximate this ideal. This paper discusses one such model system: rows of metal atoms on the Si(001) surface. These rows satisfy the requirement that they are truly one atom wide and one atom high. However, they fail in some other respects to be ideal atomic wires. Nevertheless, it is interesting to consider what is understood about their properties, how they grow, and how they might be studied further.

One can grow these wires by simply depositing a small fraction of a monolayer of certain metals onto the clean Si(001) surface at room temperature, under ultra high vacuum conditions. One monolayer is defined by the surface atomic density of Si(001). [1 ML = 6.78×10^{14} atoms/cm^2] The earliest of this

C. Joachim and S. Roth (eds.), Atomic and Molecular Wires, 11-21.

class of adsorbate metals studied by scanning tunneling microscopy (STM) was Ga.[1] This work was prompted by earlier studies that suggested that the diffusion of Ga on the Si(001) surface might be anisotropic,[2] and that the Ga formed various ordered structures for coverages up to 0.5 M, comprised of different arrangements of metal adatom dimers.[3] What the STM data showed was that at room temperature, Ga formed rows of adatom dimers on the surface, and that these rows ran perpendicular to the underlying Si dimer row direction of the Si(001) surface. As the surface coverage of Ga was increased, these rows would be crowded together, with the maximum packing density of Ga being achieved at 0.5 ML. At this coverage, the Ga adatom dimers formed an ordered 2×2 phase in accordance with earlier work.[3] The region of greatest interest for growing isolated metal rows is below this saturation coverage.

Similar behavior for room temperature growth up to 0.5 ML has now been seen for other adsorbate metals, including In,[4] Al,[5] Sn,[6] and Pb.[7] *Figure 1* shows a typical view of this low coverage behavior, in this case for 0.09 ML of In on Si(001), where the bright white lines are rows of In dimers running perpendicular to the substrate Si dimer row direction which is visible as a corrugation of the gray background. There is a single atomic height step running across the imaged area that rotates all features by 90°.

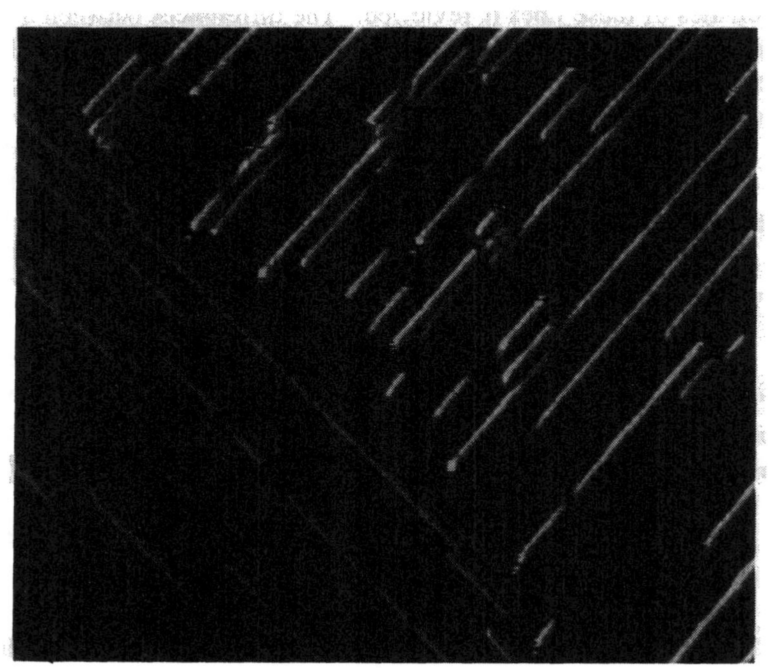

Figure 1. An empty states image (500Å × 425Å) of In on Si(001). Several very long In chains are nucleated at the lower side of the step edge.

2. Atomic and Electronic Structure: the Parallel Dimer Model

The atomic structure of these metal adatom rows is shown in *Figure 2*. This structure is called the parallel dimer model, referring to the fact that the metal atoms are dimerized in a direction parallel to the underlying Si dimers. This model was first suggested as a possibility for Al on Si(001),[5] whereas earlier work assumed the model of Bourguignon et al.[3] where the adatom dimers lie perpendicular to the underlying Si dimers. The parallel dimer model has subsequently been supported by both theory and experiment for several metal adatoms, as summarized in Table 1.

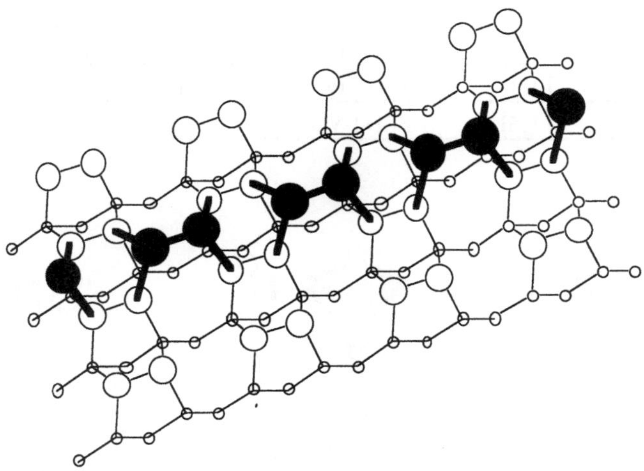

Figure 2. The parallel dimer model. Dark and white circles denote metal and Si atoms, respectively.

It is interesting that this arrangement is stable for both group III and group IV adatoms. As shown in *Figure 3*, each metal atom is bonded to two Si atoms and one metal neighbour, leaving a dangling bond at the ends of the adatom dimers. In the simplest picture, for group III metals these dangling bonds are empty, whereas for group IV metals they are half full. This difference in electronic structure has a strong effect on the STM images of these lines. For In, the metal lines are 2.0 - 2.4 Å high in empty states images (depending on the magnitude of the bias voltage) and only 0.3 - 0.5 Å high in filled states. This agrees with the significant asymmetry between empty and filled states expected from *Figure 3*. In comparison, Sn rows are roughly the same height, 1.5 - 2.0 Å, in both filled and empty states. The geometric height of these lines is expected to be of the order of the single step height on Si(001) which is 1.3 Å.

Table 1: Experimental and Theoretical support
for the parallel dimer model

Adatom	Ref.	Result
Al,Ga, In	[8]	total energy calculation for 2×2 phase, some ARUPS for double domain Al 2×2
Al	[9]	MD calculation for 2×2 structure
In	[10]	Ion scattering on single domain 2×2 surface
Al	[11]	Calculated structure, variable coverage, energy surface for Al, STM images
Al	[12]	STM images with dimer resolution
Al	[13]	ARUPS on single domain 2×2 surface
Ga	[14]	XSW and calculations for 2×2 surface
Pb	[7]	STM of Pb chains, 2×2 phase at 0.5 ML
Ga	[15]	LEED I-V analysis of 2×2 surface

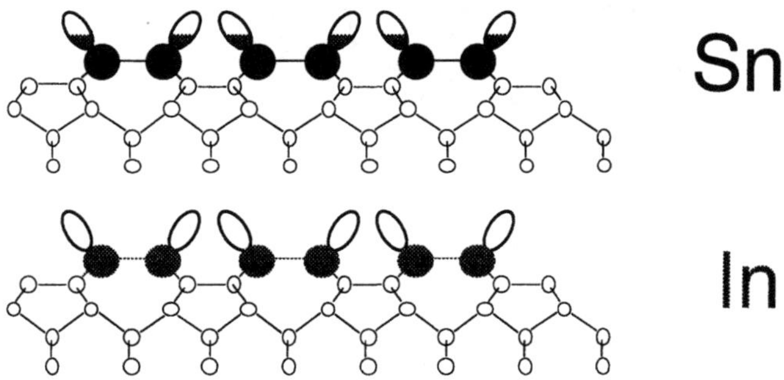

Figure 3. A side view of the parallel dimer structure for both group III and group IV adsorbates.

Electronic structure also affects the detailed atomic structure seen in the STM images of these lines, and it is important to have a comparison with theory especially in cases where the maxima in the images do not correspond to atomic positions. A high resolution filled states image of In lines on Si(001) is shown in *Figure 4*. The appearance of the In is in rough agreement with calculated images of Brocks et al for Al;[11]looking along the In row, there are broad maxima centered over the Si dimer positions, and small round maxima centered over the trenches. However, this image is different than STM images of Al on Si(001) reported by Itoh et al. where filled states maxima appear to show atomic positions.[12]

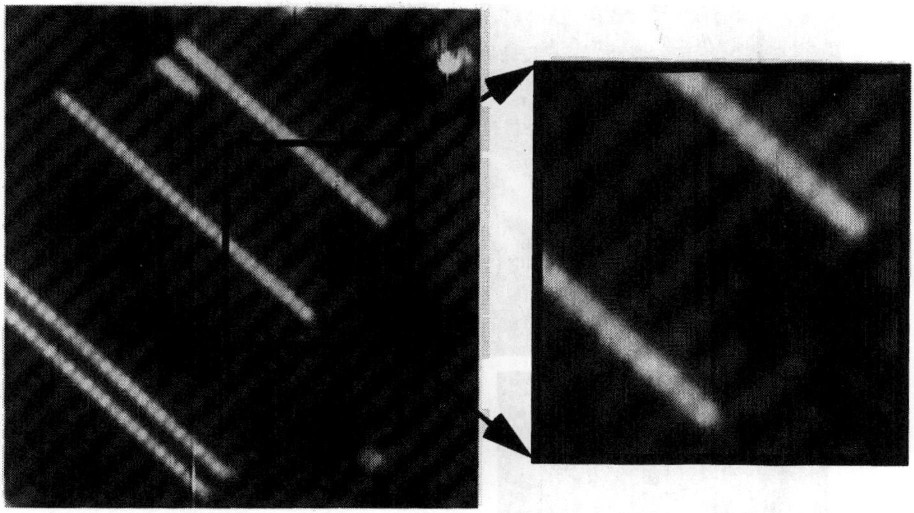

Figure 4. A high resolution filled states image (150Å × 150Å) of In lines on Si(001). The inset shows greater detail.

For Sn on Si(001) the detailed atomic structure seen by STM is more complicated. *Figure 5* shows both filled and empty states images of 0.06 ML of Sn. The empty states show more detail with maxima elongated along the Sn row direction that are strongly suggestive of parallel dimers. In the filled states, there are rounded maxima centered over the trenches between Si dimer rows.

There is also a disordering phenomenon for Sn that is not present for Al, Ga, or In. When neighbouring Sn rows are spaced the minimum allowed spacing of $2a$ apart [a = 3.84 Å = 1×1 surface unit cell length for Si(001)], the maxima can shift along the row direction so that they are not in a fixed relationship with respect to the substrate.[6] High resolution empty states data appear to show tilting of the Sn dimers in different directions along the rows. A tilting of the dimers would result in shifting of the charge density along the row, resulting in the disorder visible in the STM images. This tilting might be expected as a Peierls distortion of a chain with half filled dangling bonds in order to lower the total energy. The fact that Sn is also a comparatively large atom to fit in the parallel dimer structure might also be important.

The disordering is not present in Sn chains that are more than $2a$ apart implying that it is promoted by a short range row-row interaction that affects the atomic structure. There is currently no theory to verify the parallel dimer model for Sn, let alone one to explain the nature of this distortion and the fact that it is promoted in closely spaced chains, and not in isolated chains.

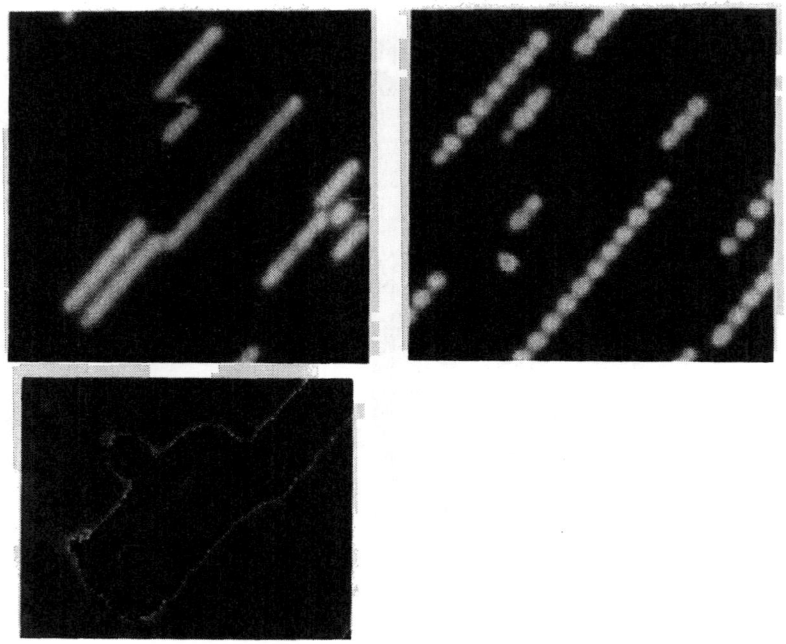

Figure 5. Empty (left panels) and filled states (right panel) images of 0.06 ML of Sn on Si(001). The lower panel is a magnified view of the filled states image, with the greyscale offset downward around the Sn rows in order to clarify the shapes of the maxima..

3. Coadsorption Studies

Brocks et al. suggest a specific reason for the strongly anisotropic growth of these metal rows: that the sites at the end of each row are strongly favoured sites for further adsorption due to their unusual bonding arrangement.[11] In order to explore further the question of how these rows grow, we studied the sequential deposition of In and Sn onto the same Si(001) surface at low coverages. The difference in electronic structure of the two species allows us to distinguish between the first and the second metal to be deposited.

Figure 6 shows 0.06 ML of Sn deposited onto 0.04 ML of In. This is a filled states image so the Sn is extremely bright, whereas the In is only slightly brighter than the surrounding Si. The appearance of the In is the same as in *Figure 4*, and most of the bright Sn maxima represent a Sn dimer. It is apparent that the Sn adsorbs in three different ways: (a) it can form rows completely independently of the In, (b) it can adsorb onto the ends of existing In rows, or (c) it can appear in the middle of a row of In.

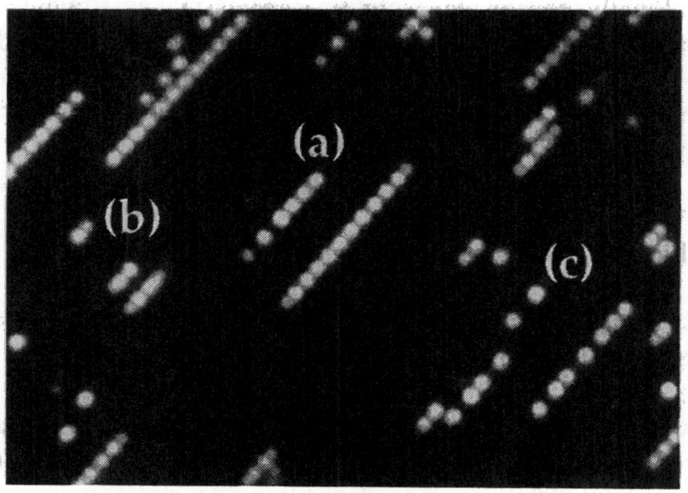

Figure 6. Co-adsorption of Sn and In.

Both (a) and (b) are consistent with the picture of Brocks et al.[11] However, it is not clear how Sn can be incorporated into the middle of an In row. There are several possibilities. One is that such incorporation only happens when Sn atoms actually land on an In row during deposition. Measurements at very low coverages of both In and Sn show more mixing than would be expected from one metal atom landing on another. The other possibilities are that metal atoms can be incorporated while diffusing up to the side of the row, or that the Sn deposition actually causes significant rearrangement of the In, so that In rows might re-nucleate in association with the Sn. In either of these cases, some rearrangement or collective motion of existing In is necessary to achieve the observed final configuration.

There are two more observations to be made from these coadsorption results. One is that the height of some of the Sn maxima that are in the middle of an In row is slightly less than that of Sn maxima attached to the end of In rows or in isolated Sn rows. This might be evidence for the formation of Sn-In heterodimers resulting from the incorporation of single Sn atoms. The second point is that the order of deposition is reversed, i.e. In deposited onto Sn, there is less mixing of the two species and more nucleation of independent rows or growth at the ends of existing rows.

The coadsorption behavior is not assumed to be the same as the behavior of two sequential depositions of one specie on the surface. Firstly, the behaviour of pure Sn and In differ in their growth and nucleation behaviour. Secondly, the difference between In on Sn and Sn on In makes it clear that there is some asymmetry in the interaction between the two species together with the substrate. Finally, the existence of two species makes strain relief, inter adsorbate chemical interaction, and entropy all factors to be considered over and above the single species growth problem. Nevertheless, the complicated behaviour seen in the coadsorption suggests that several growth mechanisms may play a role in the growth of these metal rows.

4. Metal Lines as Atomic Wires

In order to consider the properties of these rows of metal atoms as wires, we must address two issues: how long they can be grown, and whether or not they are metallic.

These rows of metal atoms self-assemble during growth, and it is not possible to control their length directly. One can study the average chain length as a function of total metal coverage, deposition rate, and growth temperature, in order to understand the growth process in more detail, and perhaps to optimize growth conditions for maximum chain length. Variations in deposition rate over more than an order of magnitude are difficult given our present apparatus. Typical growth rates used are 0.05 ML - 0.3 ML/min. and there are no observable changes for growth rates that differ by as much as a factor of two or three. Variations in temperature are limited by the fact that many of these adsorbate metals react with the Si surface at as low as 100° C. However, one can study the variation in chain length with metal coverage .

Figure 7 shows the average row length for both In and Sn as a function of metal coverage. It is clear that In grows longer rows than Sn at all coverages. The average row length increases slowly with coverage, but beyond 0.2 ML a significant number of rows are adjacent to other rows, forming small areas of the 2D 2×2 phase, and so their properties might differ from the strictly 1D isolated chains at lower coverage. As far as the longest chain lengths observed, for 0.1 ML of In, about 10% of the chains are above 40 dimers in length, i.e. longer than 300 Å. It is possible that longer continuous chain lengths may be achieved by selective nucleation of metal at parallel step edges on vicinal Si(001) surfaces.

As to whether or not these chains are metallic, it is fairly likely that the group III metal chains would be semiconducting given that the 2×2 surface is semiconducting. The group IV chains are more likely to be metallic. *Figure 8* shows an average scanning tunneling spectroscopy (STS) spectrum on the Sn chains, along with a curve for clean Si, with both spectra derived from data in different areas of the same image. The curves for clean Si are in good agreement with the literature.[16] The curves for the Sn chains are not metallic, and they appear to show a band gap that is slightly larger than that of the Si(001). Nevertheless, it is not obvious that STS data reflects the conductivity of the chains along the surface as the STM measures conductivity perpendicular to the surface. Some theoretical work will be necessary to understand whether the group IV chains are conductive.

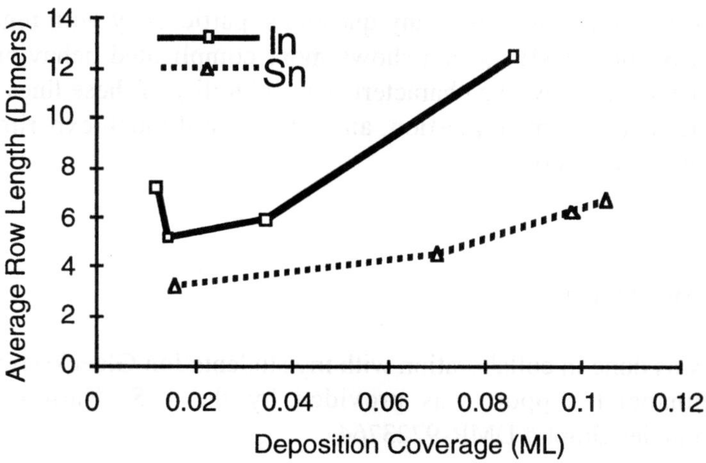

Figure 7. Average row length versus metal coverage for In and Sn.

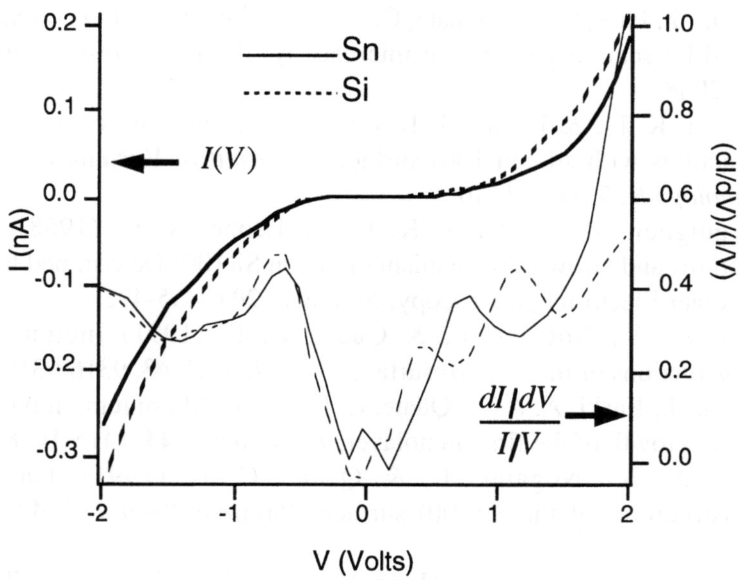

Figure 8. STS curves for Sn on Si(001). The upper curves are I(V) and the lower curves are the logarithmic derivative.

5. Summary

In summary, group III and group IV metals are known to self-assemble into single atom wide lines on the Si(001) surface. STM images along with other experimental data and theory help us to understand the basic structure of these lines. However, there are still many questions, particularly with regard to the behavior of Sn on Si(001) which shows more complicated behavior than the group III metals. Finally, the characteristics and utility of these lines as atomic wires also remains an open question, and there is still much experimental and theoretical work to be done.

6. Acknowledgements

This work was done in collaboration with my students Jon Glueckstein and Matt Evans. Financial support was provided by the U.S. National Science Foundation under Grant # DMR-9223764.

7. References

1. Nogami, J., Park, S.-I., & Quate, C. F. (1988) Behavior of Ga on Si(100) as studied by scanning tunneling microscopy, *Applied Physics Letters*, **53**, 2086-2088.
2. Carleton, K. L., & Leone, S. R. (1987) Laser probing of gallium atom interactions with silicon(100) surfaces, *Journal of Vacuum Science and Technology B*, **5**, 1141-1146.
3. Bourguignon, B., Carleton, K. L., & Leone, S. R. (1988) Surface Structures and Growth Mechanism of Ga on Si(100) Determined by LEED and Auger Electron Spectroscopy, *Surf. Sci.*, **204**, 455-472.
4. Baski, A. A., Nogami, J., & Quate, C. F. (1991) Indium-induced reconstructions of the Si(100) surface, *Phys. Rev. B*, **43**, 9316-9319.
5. Nogami, J., Baski, A. A., & Quate, C. F. (1991) Aluminum on the Si(100) surface: growth of the first monolayer, *Phys. Rev. B* **44**, 1415-1418.
6. Baski, A. A., Nogami, J., & Quate, C. F. (1991) Tin-induced reconstructions of the Si(100) surface, *Physical Review B*, **44**, 11167-11177.
7. Li, L., Koziol, C., Wurm, K., Hong, Y., Bauer, E., & Tsong, I. S. T. (1994) Surface morphology of Pb overlayers grown on Si(100)-(2×1), *Physical Review B*, **50**, 10834-10842.

8. Northrup, J. E., Schabel, M. C., Karlsson, C. J., & Uhrberg, R. I. G. (1991) Structure of low-coverage phases of Al, Ga, and In on Si(001), *Physical Review B*, **44**, 13799-13802.
9. Adams, G. B., & Sankey, O. F. (1992) Applications of Ab Initio Quantum Molecular Dynamical Relaxation: Silicon(111)-5×5 Surface Reconstruction and Aluminum deposited on Si(100), *Journal of Vacuum Science and Technology A*, **10**, 2046.
10. Steele, B. E., Li, L., Stevens, J. L., & Tsong, I. S. T. (1993) Structure of the Si(100)-(2×2) In surface, *Physical Review B*, **47**, 9925-9927.
11. Brocks, G., Kelly, P. J., & Car, R. (1993) Adsorption of Al on Si(100): A Surface Polymerization Reaction, *Physical Review Letters*, **70**, 2786-2789.
12. Itoh, H., Itoh, J., Schmid, A., & Ichinokawa, T. (1993) Structures of low-coverage phases of Al on the Si(100) surface observed by scanning tunneling microscopy, *Physical Review B*, **48**, 14663-14666.
13. Yeom, H. W., Abukawa, T., Takakuwa, Y., Nakamura, M., Kimura, M., Kakizaki, A., & Kono, S. (1994) Surface electronic structure of single-domain Si(001)2×2-Al: an angle-resolved photoelectron spectroscopy study using synchrotron radiation, *Surface Science Letters*, **321**, L177-L182.
14. Qian, Y., Bedzyk, M. J., Tang, S., Freeman, A. J., & Franklin, G. E. (1994) Resolving the Ga Ad-Dimer Location and Orientation on the Si(100) Surface, *Physical Review Letters*, **73**, 1521-1524.
15. Sakama, H., Murakami, K., Nishikata, K., & Kawazu, A. (1994) Structure of a Si(100)2×2-Ga Surface, *Physical Review B*, **50**, 14977-14982.
16. Hamers, R. J., Avouris, P., & Bozso, F. (1987) Imaging of Chemical-Bond Formation with the Scanning Tunneling Microscope: NH_3 Dissociation on Si(001), *Physical Review Letters*, **59**, 2071-2074.

ATOMIC DEPASSIVATION AND METALLIZATION OF THE Si(100)-2x1:H SURFACE

T.-C. SHEN, C. WANG, and J. R. TUCKER
Department of Electrical & Computer Engineering
University of Illinois at Urbana-Champaign
Beckman Institute, 405 North Mathews Avenue
Urbana, IL 61801, USA

1. Introduction

Atomic resolution has recently been achieved in STM-induced desorption of individual hydrogen atoms from silicon dangling bonds on *in situ* H-passivated Si(100)-2x1 surfaces.[1] The bare silicon orbitals thus created can potentially be reacted with SiH_4, NH_3, AsH_3, B_2H_6, and a wide variety of other gases to selectively epitax, nitride, and dope the silicon surface in areas no longer protected by the monolayer hydrogen resist. Thus far, we have selectively reacted the STM-exposed regions with pure O_2 to create oxide patterns having ~30Å minimum feature size,[2] and these ultra-thin oxides are currently being used as masks for a variety of pattern transfer processes. Here we describe initial experiments in which aluminum atoms are evaporated directly onto patterns written by STM in the H-atom resist at room temperature, without oxidation. Because individual metal atoms diffuse freely over H-terminated portions of the surface still coated with hydrogen, they can be captured within STM-exposed regions of bare silicon under appropriate conditions. In order to utilize this effect to create atom-resolved metal patterns on silicon, the evaporation rate must kept be low and the surface diffusion high enough to permit capture of individual metal atoms in regions of bare silicon before they undergo collisions. Collisions between metal atoms on the H-terminated areas result in formation of metal "nanoclusters" which become immobilized due to their increased mass. In the initial experiments reported here, most aluminum atoms reach STM-exposed areas of bare silicon in patterns of ~20Å-wide lines on 60Å-pitch. By using very low deposition rates and maximizing diffusion of metal atoms over the H-termination, it appears that this technique could be refined to fabricate single-atom wires and devices in a form suitable for electrical measurement, offering extraordinary new opportunities to produce metallic structures at the ultimate limits imposed by nature.

2. Sample Preparation

Atomically clean silicon surfaces are prepared in ultra-high vacuum (UHV) with base pressure ~$5x10^{-11}$ Torr by heating to 1250°C for approximately 1 minute. The chamber is then backfilled with 10^{-6} Torr of H_2, and a tungsten filament at ~1500°C is placed 6cm from the surface to generate atomic hydrogen. The sample is restively heated to ~350°C

C. Joachim and S. Roth (eds.), Atomic and Molecular Wires, 23-29.

during the hydrogen exposure in order to prevent formation of higher hydrides and etching, and a nominal dose in excess of 400L [1L=10^{-6} Torr s] is employed to ensure saturation of all surface dangling bonds. Under these conditions, the 2x1 dimer reconstruction of Si(100) is preserved with the addition of a single hydrogen to each surface silicon atom.

Patterns are written in this monolayer H-atom resist by STM-induced desorption of hydrogen atoms using the highly localized low-energy electron beam of a scanning tunneling microscope. Figure 1 shows 30Å-pitch lines written along the dimer row direction. Every silicon dimer is visible in this occupied states image. The bright ball-like features are bare silicon dangling bonds where a lone hydrogen atom has been desorbed, and thicker portions of these lines mark regions where both hydrogens are eliminated from paired silicon atoms within a single dimer row.

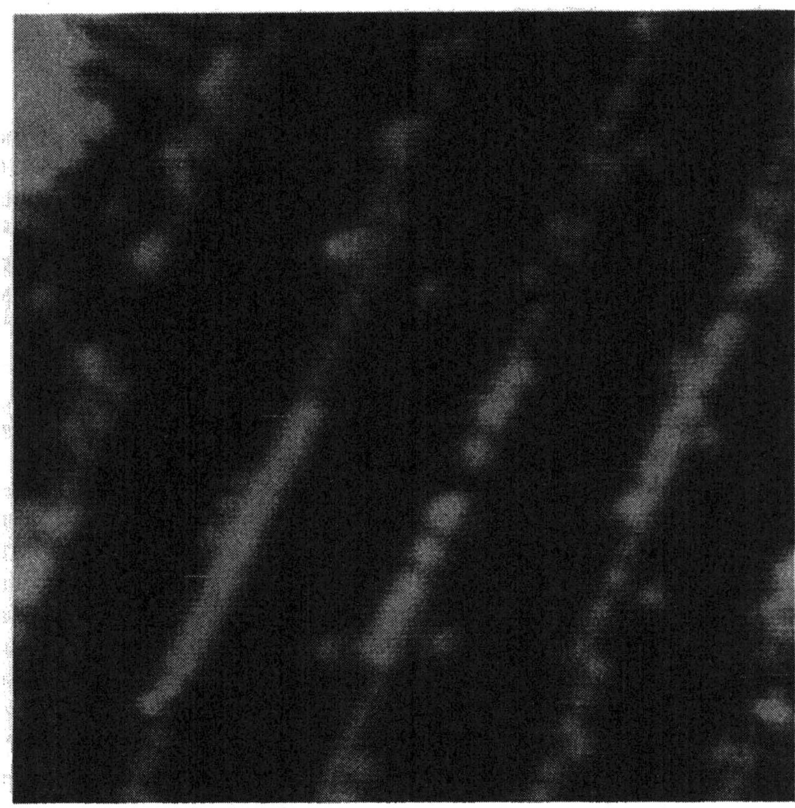

Fig. 1 UHV-STM $(150Å)^2$ image showing 30Å-pitch lines of bare dangling bonds created by STM-induced desorption of individual hydrogen atoms on an H-terminated Si(100)-2x1 surface. Writing conditions here are tip bias –3V, current 4.5nA, and line dose $5x10^{-4}$ C/cm. Imaging conditions are +1.4V, 0.1nA.

Figure 2 shows a $(2400Å)^2$ STM image containing a larger pattern written by raising the tip voltage to –4V and the demanded current to 4nA. The five lines in the center are ~20Å wide on 60Å-pitch and were written using a line dose of $2x10^{-3}$C/cm, corresponding to a scan speed of 200Å/s. The two 100Å-wide outer lines were made by superposing 9 such lines with 10Å-pitch. The STM-desorbed areas appear bright under nonperturbative imaging conditions, here –2V and 0.1nA, because tunneling into bare silicon dangling bonds is substantially enhanced compared to regions still covered with hydrogen atom resist. High resolution images show that STM-exposed areas maintain continuity in 2x1 dimer reconstruction across their boundaries to H-terminated portions of the surface. Typical roughness of these boundaries between patterned and unpatterned areas is approximately one dimer row.

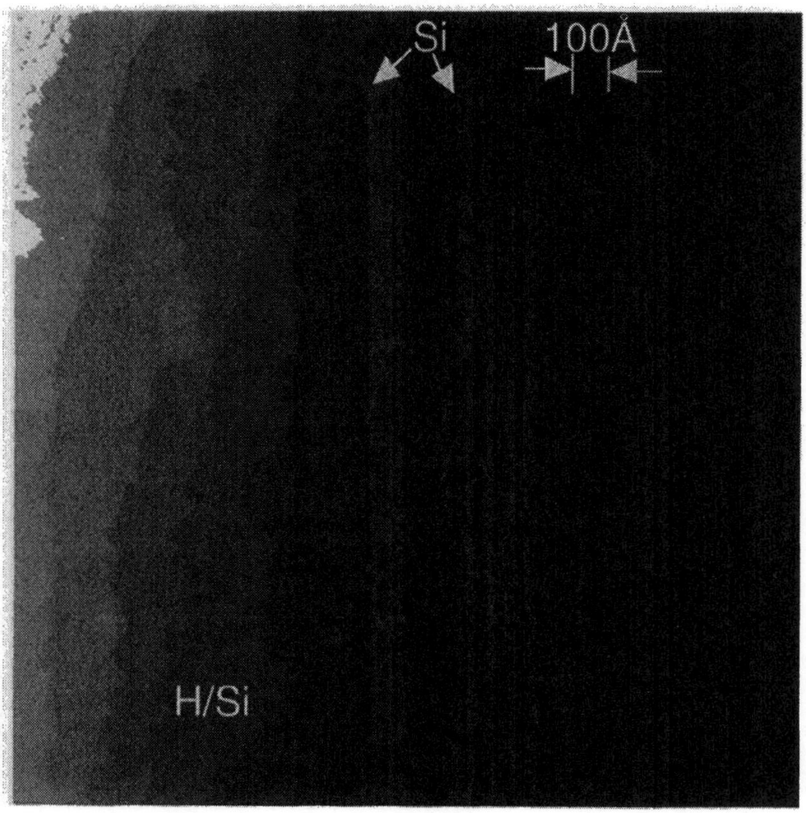

Fig. 2 UVH-STM image $(2400Å)^2$ showing lines of bare silicon written by STM-induced desorption of hydrogen atoms on an H-terminated Si(100)-2x1 surface, employing tip bias –4V, current 4nA, and line dose $2x10^{-3}$C/cm.

3. Aluminum Deposition

Figure 3 shows the same pattern after evaporation of 0.1ML aluminum onto the sample surface at room temperature. The chamber pressure was kept $< 5\times10^{-10}$ Torr during the evaporation. In this $(2400Å)^2$ image, very small ball-like features are clearly observed on the H-passivated portions of the surface following aluminum deposition. A close-up view of one such H-terminated area is shown in Fig. 4. Here the Al nanoclusters can be seen in relation to silicon dimer rows. Their size has a continuous distribution, observed in many such images, from a minimum of 2 atoms up to a well defined maximum. The average density of Al clusters on the H-terminated surface is found to be 3×10^{12} cm^{-2} at 0.1ML deposition under these conditions, corresponding to an average of ~20 aluminum atoms per cluster.

Fig. 3 Result of evaporating 0.1ML aluminum onto the STM-patterned H-terminated Si(100)-2x1 surface shown in Fig. 2. Individual Al atoms diffuse and collide on H-terminated portions of the surface to form immobilized aluminum nanoclusters, with an average of ~20 metal atoms per cluster under these conditions.

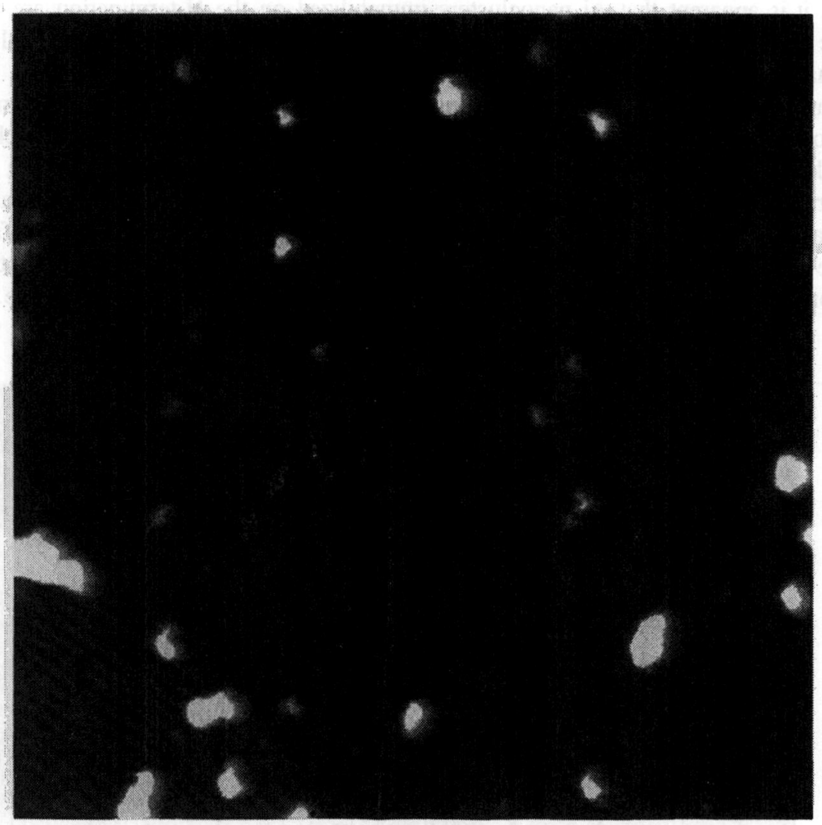

Fig. 4. STM image (400Å)^2 of Al nanoclusters on a portion of the H-terminated surface in Fig. 2, following deposition of 0.1ML aluminum. Area density of Al islands is 3x10^{12}cm^{-2} on the H-termination with an average of ~20 metal atoms per cluster.

Purity of the metal deposition in these experiments was verified by observing the formation of individual Al-chains transverse to the silicon dimer rows for low coverages on clean Si(100)-2x1 surfaces without hydrogen passivation, and an ordered 2x2 reconstruction for full coverage.[3,4] Thickness monitor calibration was also checked by atom-counting.

Individual Al adatoms appear to diffuse much more easily on H-terminated portions of the silicon surface than on bare silicon, and it may be possible to take advantage of this effect as a means for selective metal deposition within STM-patterned areas. Figure 5 is an (800Å)^2 image showing a portion of the STM-desorbed pattern in Figs. 2 and 3 following deposition of 0.1ML aluminum at room temperature. On this scale, it is apparent that Al nanoclusters immobilized on the H-termination are far less dense between the 60Å-pitch lines than elsewhere, indicating that most Al atoms originally deposited within these regions have diffused into STM-exposed areas of bare silicon without suffering a collision. While metal patterning in these initial experiments is not completely selective even in such finely patterned areas, greatly enhanced selectivity can likely be realized by employing very low deposition rates in combination with modest substrate heating to enhance surface diffusion of metal atoms over the H-termination. Similar experiments by Wada, et al.[5] employ evaporated Ga atoms an effort to fabricate single-atom wires. With persistence, it should eventually be possible to approach the goal of atomically defined metal structures on H-passivated silicon using this technique.

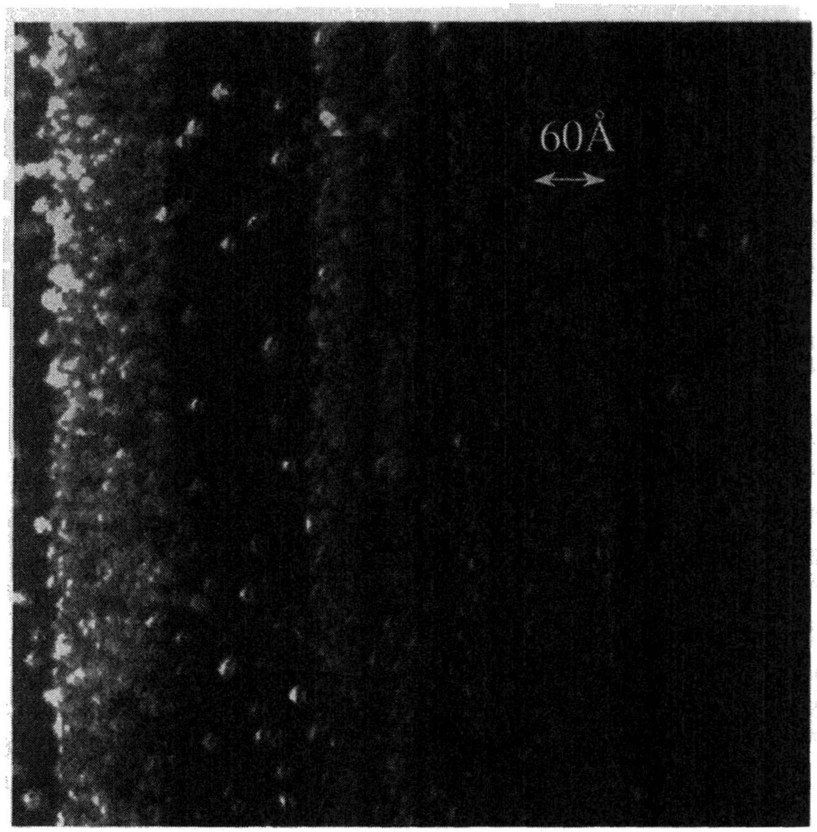

Fig. 5 STM image $(800\text{Å})^2$ of bare silicon lines in Figs. 2 and 3 following deposition of 0.1ML aluminum. The density of Al nanoclusters on H-terminated areas lying between the 60Å-pitch lines is much smaller than elsewhere, indicating that most Al atoms landing in these regions have diffused to STM-exposed lines of bare silicon.

4. Conclusion

In these experiments, we have shown that individual Al atoms diffuse far more easily over H-terminated Si(100)-2x1 surfaces than on bare silicon. Because STM-desorption of hydrogen can be performed with atomic precision to expose bare silicon dangling bonds, the possibility exists for patterning of metal structures on the atomic scale. Through selection of metal species and substrate temperature for maximum diffusion over H-terminated portions of the surface and use of very low deposition rates, it should be possible to approach complete selectivity in metal deposition onto STM-generated nanoscale patterns. While small structures comprising several atoms can be assembled through atom manipulation, alternative methods must be sought in order to fabricate extended metal wires and devices for electrical measurement. An approach based upon STM-induced hydrogen desorption to create templates on H-terminated silicon now appears to offer unique potential for realizing the ultimate limit of resolution in defining such metal structures on the atomic scale.

Acknowledgements

This work is supported by the U.S. Office of Naval Research under grant N00014-92-J-1519 "STM-Based Nanolithography"

References

1. T.-C. Shen, C. Wang, G. C. Abeln, J. R. Tucker, J. W. Lyding, Ph. Avouris, and R. E. Walkup (1995), Atomic scale desorption using electronic and vibrational excitation mechanisms, Science **268**, 1590-1592.

2. T.-C. Shen, C. Wang, J. W. Lyding, and J. R. Tucker (1995), Nanoscale oxide patterns on Si(100) surfaces, Appl. Phys. Lett. **66**, 976-978.

3. M. A. Hasan, G. Radnoczi, J. E. Sundgren, and G. V. Hansson (1990), Epitaxial growth of Al on Si by thermal evaporation in ultra-high vacuum: Growth on Si(100)2x1 single and double domain surfaces at room temperature, *Surf. Sci.* **236**, 53.

4. J. Nogami, A. A. Baski, and C. F. Quate (1991), Aluminum on the Si(100) surface: Growth of the first monolayer, *Phys. Rev.* **B44**, 1445.

5. Y. Wada, et al., these proceedings.

CONTROLLED MANIPULATION OF ATOMS AND MOLECULES AND FORMATION OF NANOSTRUCTURES WITH THE SCANNING TUNNELING MICROSCOPE

GERHARD MEYER, SVEN ZÖPHEL, AND KARL-HEINZ RIEDER
Institut für Experimentalphysik, Freie Universität Berlin,
Arnimallee 14, D-14195 Berlin, Germany

Abstract

The scanning tunneling microscope is not only suited for imaging surfaces with atomic resolution, but has been developed recently to an operational instrument able to manipulate single atoms and molecules in a controlled manner. Here we present several examples of manipulation on atomic scale and formation of nanostructres with these techniques. Lateral manipulation, i.e. sliding of isolated species along the surface, is demonstrated for CO, C_2H_4, Pb-atoms, and Pb-dimers on a Cu(211) substrate. Vertical manipulation, i.e. controlled transfer of species between tip and surface and vice versa is demonstrated with Xe and C_3H_6. The design of our low cost, versatile low temperature scanning tunneling microscope capable of manipulations on atomic scale is described briefly.

Introduction

Since its invention in 1983 the scanning tunneling microscope (STM) has been established as a powerful tool in surface science [1]. Apart from its capability to image surfaces with atomic resolution it was shown that the STM is able to manipulate single atoms and molecules with atomic scale precision [2]. To do this, the different physical and chemical interaction mechanisms between tip and sample are employed. One technique is to solely use the forces that act on the sample because of the proximity of the tip. This means that long range van der Waals and short range chemical forces can be used to slide atoms or molecules along the surface. No bias voltage or tunneling current are necessary in principle. We will concentrate on this kind of "soft" manipulation in the following. Further techniques using the effects caused by the electric field generated by the bias voltage between tip and sample and by the current flowing through the gap region are used for atomic or molecular modification [3]. Due to the weakness of the van der Waals forces the first experiments on "soft" manipulation of individual atoms and small molecules have been done at 4 K [2]. Meanwhile also one experiment involving the sliding of large organic molecules has been performed at room temperature [4]. Our experiments concern the manipulation of atoms and small molecules at temperatures between 15 and 80 K [5]. We will start with a description of our low cost

C. Joachim and S. Roth (eds.), Atomic and Molecular Wires, 31-45.

versatile low temperature STM used in the experiments on atomic manipulation. Two kinds of manipulation will be discussed, the first being the lateral manipulation of atoms and molecules, i.e. sliding of atoms and molecules along the surface. During the sliding process the manipulated species are always bonded to the surface. As examples we discuss CO, C_2H_4 and Pb on Cu(211). The second technique deals with the vertical manipulation of atoms and molecules, i.e. the transfer of atoms and molecules between the sample surface and the tip and vice versa. As examples we discuss Xe and C_3H_6.

1. Experimental

Our low temperature scanning tunneling microscope was originally designed as a modular setup to allow for simultaneous STM and AFM operation in UHV. Here we discuss only the STM part of the microscope. The scanner (Fig. 1) is similar to the well known design principle of the "Besocke beetle type scanner" [6, 7] which exhibits inherent high mechanical stability and good drift compensation. For temperature measurement a silicon diode sensor [Lake Shore DT-471-SD] is mounted on top of the sample ring and another on the base plate of the scanner. The sample, which is mounted in a special molybdenum sample holder, is positioned on top of the sample ring and is resting there simply due to gravity. The difference between the sample temperature and the sensor temperature was measured to be less than 5 K in the main operating temperature range between 15 and 150 K. With a resistance heater the sample ring can be slowly heated up close to room temperature (no tunneling experiments are performed during the actual heating up process due to drift). To achieve good mechanical vibration isolation the microscope is hanging from helical stainless steel springs, which are attached to the lower part of a liquid helium bath cryostat (Fig. 1). Three magnets at the bottom of the baseplate of the microscope are used for eddy current damping. The whole STM is surrounded by two concentric radiation shields. In tunneling position the microscope is hanging from the stainless steel springs and has no direct contact to the He bath. This results in good mechanical vibration isolation but also in a weak thermal contact to the He bath. Since the microscope is surrounded by a 4 K radiation shield, no thermal radiation is falling onto the STM, so that the STM is almost thermally isolated. The heat capacity of the STM and the thermal conductance of the wires connected to the STM determine the warming-up rate. Since the heat capacity decreases strongly with decreasing temperature reasonable temperature stability (i.e. temperature increases less than a few K/h) can be achieved for temperatures only above 30 K with this technique. On the other hand the thermal isolation allows the microscope to be heated without putting much heat load on the liquid helium bath. Cooling the STM is performed simply by pulling the STM down with a small hook, which is operated by a wobblestick, until the baseplate makes mechanical contact to the inner radiation shield (Fig. 2). At the same time the sample ring with the ramps and the scanner are separated. In this position not only cooling is achieved but there is also the possibility to transfer samples in and out of the STM. Typical cooling times between room temperature and for example 30 K are 4-5 h. To reduce the operating temperature below 30 K, copper braided wires are installed between the scanner baseplate and the inner radiation

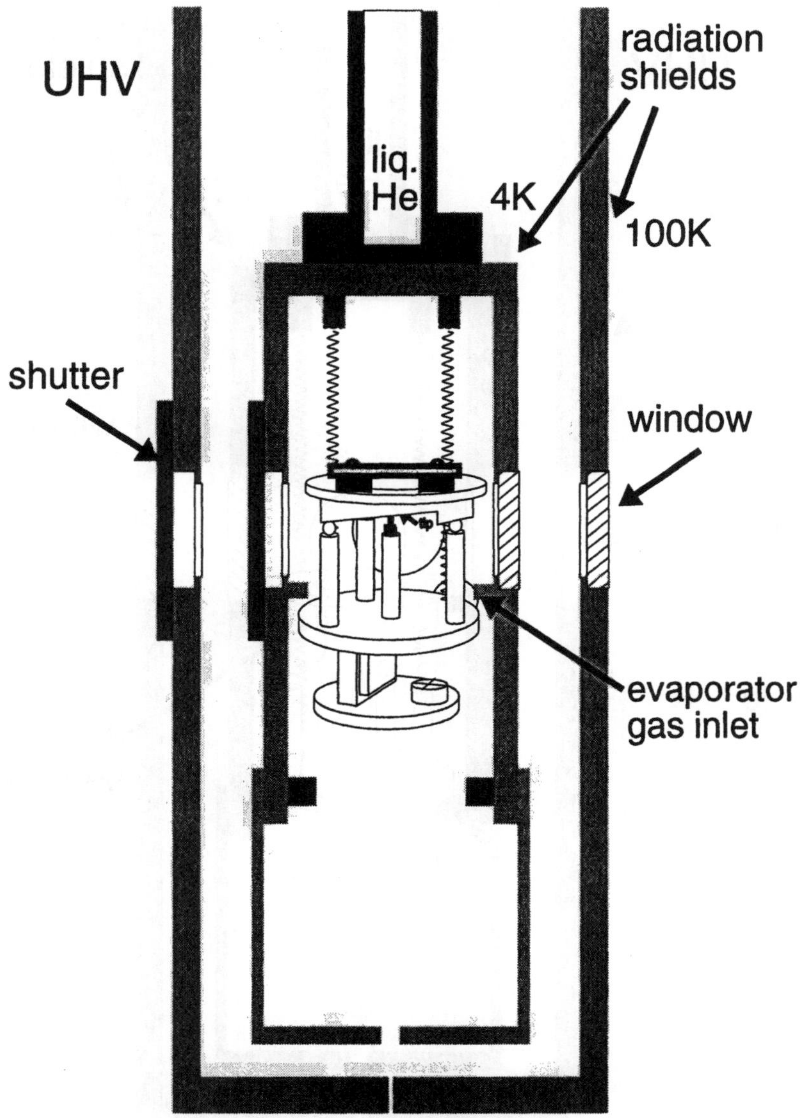

Figure 1. Overview of the microscope in the tunneling position. The scanner is hanging from springs off the bottom of the cryostat, to which also the inner radiation shield is connected. Apart from the inner radiation shield at 4K a second one at 100 - 120 K is installed. Shutters and windows are built in.

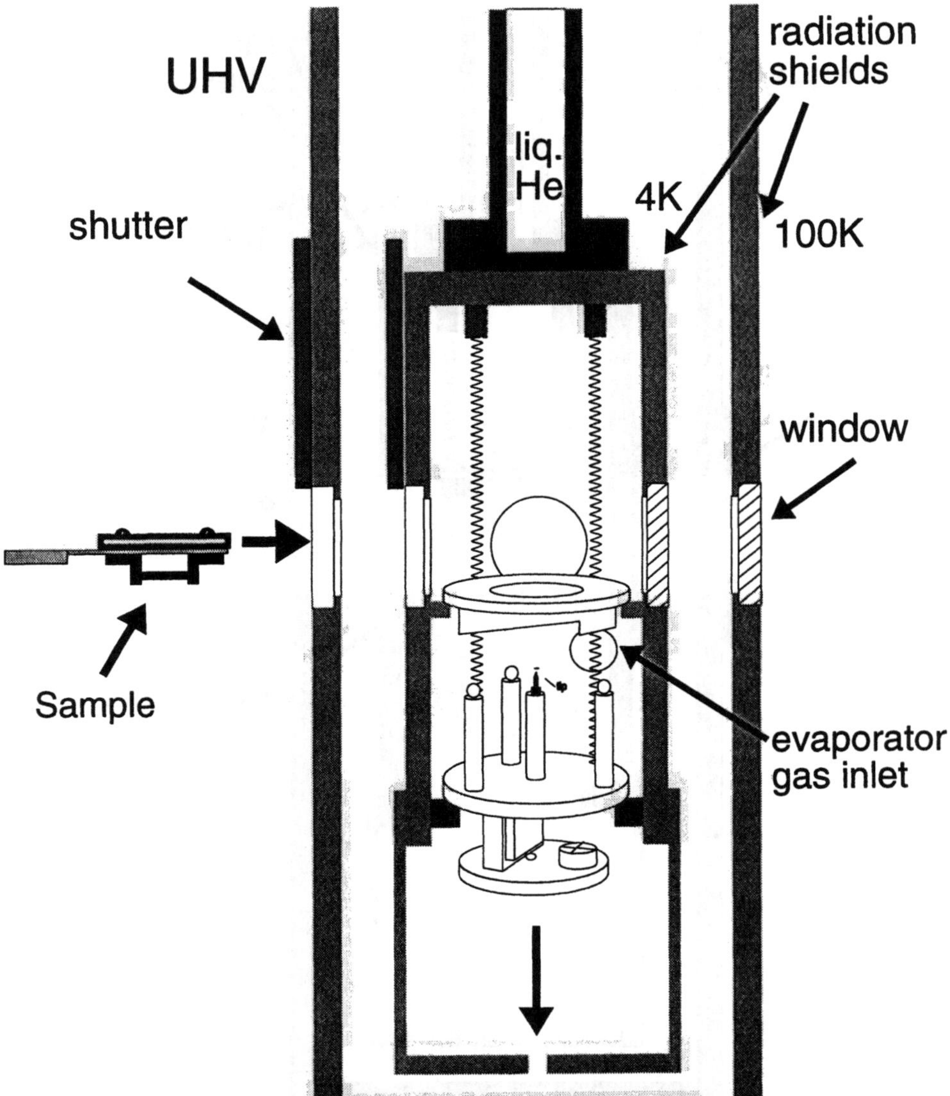

Figure 2. Cooling and sample transfer position of the microscope. Scanner and sample ring are separated and are making contact to the inner radiation shield, if the microscope is pulled down. Opening of the shutters allows sample transfer into the microscope.

shield. Because of the high thermal conductance of the copper wires the temperature of the baseplate is permanently held at 7 K and the temperature of the sample ring at 15 K. Since the installation of the copper wires decreases the mechanical vibration isolation, they are only used in the low temperature range, where liquid helium is used as coolant. Noise levels of less than 0.003 nm peak to peak are observed. The built-in windows (DURAN) allow a well controlled tip approach and the investigation of structured or very small samples with the help of an optical microscope. On the inner and outer radiation shields shutters are installed, which can be opened for sample transfer. A separate shutter on the backside of the radiation shields can be opened for gas inlet and in situ evaporation. The solid angle under which gas can adsorb onto the sample is only 1/3000 of 2 π. Therefore, a factor of 3000 higher partial pressures have to be used to reach the same dose as compared to a free unshielded sample. Very clean operation of the system is possible, because a very high 'signal to noise' ratio relative to the residual gas components is achieved. Consequently very small amounts of adsorbates can be adsorbed in a controlled manner and a once prepared sample can be studied for a long time without contamination by residual gases. This is particularly important for manipulation experiments, because in this case one usually 'works' only with a few atoms or molecules.
The electronics for control of the STM is based on a digital design allowing for flexible control and easy modification of the STM. Furthermore it simplifies the implementation of techniques like the manipulation of single atoms and molecules which is discussed in the following.

2. Lateral Manipulation

2.1 THE CLEAN CU(211) SURFACE

All experiments have been done on the stepped copper (211) surface. The topview of a sphere model of the clean Cu(211) substrate is shown in Fig. 3: The surface consists of (111) terraces separated by (100) single step facets. The dimensions of the unit cell are: $a_1 = 0.255$ and $a_2 = 0.625$ nm. We define 1 ML as one adsorbate particle per 1×1 unit cell. In all STM pictures shown the surface orientation is (apart from some tilt) the same as the one plotted in Fig. 3. The surface was prepared by several cycles of sputtering with 800 eV Ne ions and annealing to 800 K.

2.2 CARBONMONOXIDE ON CU(211)

It is well established for metal surfaces, that CO molecules at low coverages usually adsorb upright or tilted at on top positions with the carbon atoms close to the metal atoms [8]. On Cu(211), the CO molecules occupy top sites over those Cu atoms, which form the ridges of the Cu(211) substrate; the CO molecules are probably tilted against the (111) nanofacets of the (211) substrate [9].
As one of the first structures we built using lateral manipulation of CO molecules we show in Fig. 4 the letter 'X' [10] at a surface temperature of 15 K. At this temperature

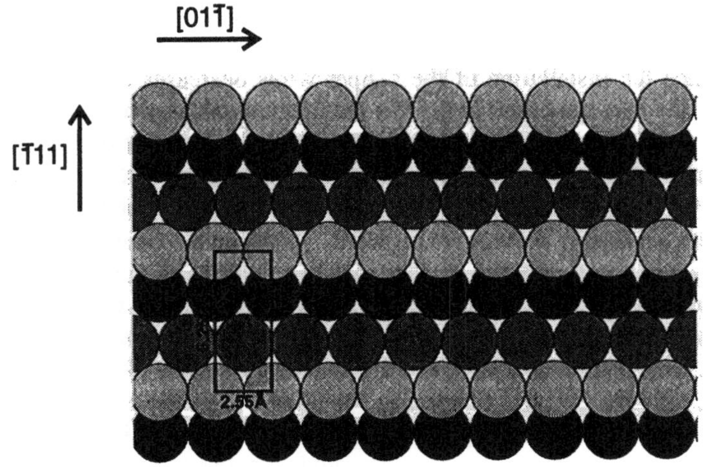

Figure 3. Sphere model of the Cu(211) substrate. The Cu atoms are shown as small spheres with the deepest lying atoms shaded darkest. The surface unit cell is indicated. The nearest neighbour distance of Cu atoms is 0.255 nm and the distance between the step edges amounts to 0.625 nm.

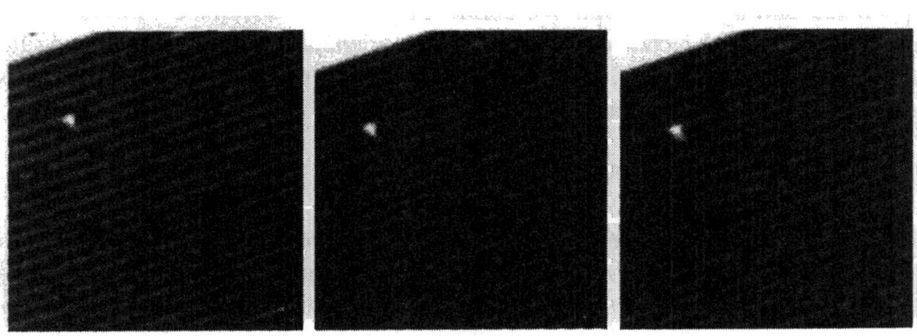

Figure 4. Letter 'X' formed by controlled lateral manipulation of CO-molecules on a Cu(211) surface. T = 15 K. Size of each image: 13.5 nm*13.5 nm.

thermal excitation leading to irregular diffusive jumps of the CO molecules is appreciably reduced as compared to 30 K, so that nanostructures can be formed with perfect precision. No changes in the structure of the letter 'X' and its surroundings were observed an hour after its build-up proving the structural stability of the nanostructures formed with the weakly chemisorbed CO-species at 15 K. This leads to the expectation, that "wires" and even structures like "crossed wires" in the nano- and subnanorange can be built up in the future with good reliability. Even more important, the fact that with our STM we scanned exactly the same area at the surface 60 minutes later is proof of the extreme mechanical stability of the instrument at these low temperatures (compare Figs. 4b and Fig. 4c. A tip change having occurred between Fig. 4a and Fig. 4b is the reason for the lateral shift between these two images). A lateral drift on the order of 0.1nm/h is typical for our system (at higher temperatures the drift is much larger.)
For manipulation of the CO molecules we use the procedure illustrated schematically in Fig. 5 [2]: Whereas in the STM imaging process the tip is scanned at distances of a few atomic diameters above the surface (a) and – in the constant current mode – follows contours of constant local electronic density of states [11, 12], for the manipulation of adsorbates the tip is brought close to the molecule (b) by reducing the tunneling resistance and then the tip is moved parallel to the surface (c) by a pregiven distance whereby the molecule is dragged (or pushed) to the desired location (d). Then the tip is drawn back to a larger distance (e) and a new STM picture is scanned to check the result of the manipulation. It is of importance to notice, that due to the pronounced anisotropy of the Cu(211) substrate, manipulation of the CO molecules along the step edges proceeds readily, whereas manipulations perpendicular to the steps are almost impossible. We can also determine from our experiments that the force on the molecule is attractive. Unfortunately the magnitude of the force acting on an atom or molecule is not directly accessible to the STM. On the other hand, the force is a function of the distance between tip and adsorbate. Controlling the distance thus gives a measure of the exerted force. Moreover, since the tunneling current at low voltages is proportional to the bias voltage and depends exponentially on the distance [11], the tunneling resistance can be used as a rough measure to control the force acting on the adsorbate [12]. This results in a threshold of the tunneling resistance below which adsorbates can be moved. The indirect control of the force results in the consequence that the tunneling resistance values depend significantly on the tip structure. Careful and reproducible tip preparation is therefore an important factor. In the case of sliding single CO molecules parallel to the step edges a reduction of the tunneling resistance to values between 600kΩ and 1MΩ was sufficient to obtain precise lateral movements.

2.3 ETHEN ON CU(211)

The second example of lateral manipulation concerns C_2H_4. The ethen molecules are located at the step edges of the Cu(211) surface. The same behaviour has been observed on Cu(110) [13]. From registration information relative to single copper atoms we can determine that C_2H_4 is located symmetrically to the two fold bridge site at the step edge [14] with the C-C axis of the molecule aligned parallel to the step edges. No rotation of the molecule relative to the step edges is observed. (In that case due to symmetry two

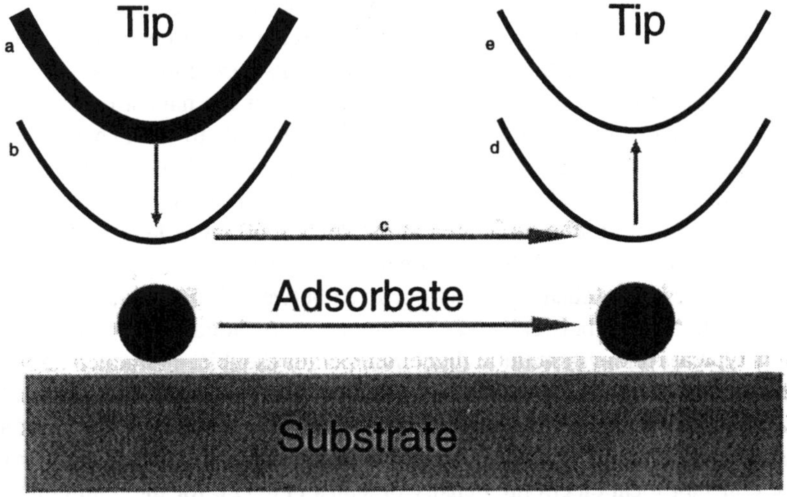

Figure 5. Principle of lateral manipulation (the details are described in the text).

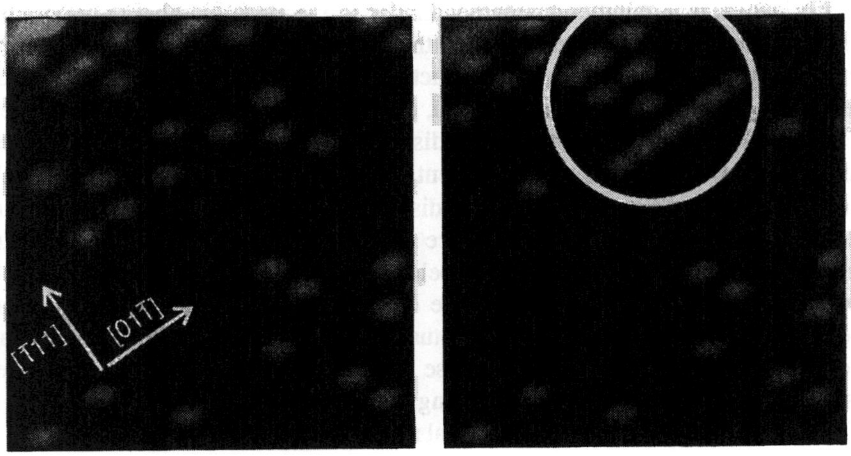

Figure 6. Lateral manipulation of C_2H_4 molecules. The STM images show the formation of the 'tip-sample' symbol inside the white circle. T =35 K. Image size: 7.5 nm * 7.5 nm.

kinds of orientations of the molecules would be expected in the images.) Lateral manipulation of C_2H_4 parallel to the step edges is possible for tunneling resistances less than about 500 kΩ. Fig. 6 shows the buildup of a nanostructure representing schematically the tunneling tip above a surface. These experiments were performed at a temperature of 40 K. Controlled manipulation perpendicular to the step edges appears possible but is much more difficult and also tip dependent.

2.4 LEAD ON CU(211)

The last example deals with the lateral manipulation of single Pb-atoms and Pb-dimers. Lead has been evaporated from a Knudsen cell at sample temperatures between 20 and 80 K. At these low temperatures Pb is adsorbed in the 5 fold sites located downwards the step edge. This can be deduced from registration relative to single copper atoms [14], which certainly are located also in the 5 fold sites down at the step edge, as these are their natural bulk positions. While on other copper surfaces the incorporation of single Pb-atoms into the substrate has been observed at room temperature [15], the temperatures used here are low enough to prevent single Pb-atoms from being incorporated into the substrate. Fig. 7 shows four Pb-atoms positioned by lateral manipulation to form the corners of a rectangle. The distance of the Pb-atoms parallel to the step edges is six next neighbour copper distances. The threshold tunneling resistance to manipulate single Pb-atoms varied between 400 - 600 kΩ, which is only slightly smaller than in the previous examples. The manipulation force is again attractive. Apart from single Pb-atoms we observe also Pb-dimers on the surface even at low coverages (see Fig. 8a). The Pb-dimers appear only slightly higher (0.005 nm) than single Pb-atoms. (This height difference is much smaller than the overall height of a Pb-atom or dimer of 0.1 nm and therefore is not easily detected in Fig. 8a). The mesh size of the grid in Fig. 8a corresponds to the size of the (211) unit cell. From this registration information we determine that the Pb-dimer is symmetrically positioned to the on top site at the step edge (Fig. 8b). The tunneling resistance for manipulation of single dimers is close to the tunneling resistance of single Pb-atoms. The dimer is also stable with respect to lateral manipulation, i.e. we did not succeed to separate a dimer by lateral manipulation. On the other side, it is also not possible to form longer Pb-chains. To demonstrate this, Fig. 9 shows the sliding of a dimer with an extra Pb-atom in front of it. The result shows that no trimer has formed; this is not surprising as no trimers or longer chains have been observed after further evaporation. It shows further that in our experiments the dimers are stable against the exchange of an atom between the dimer and the single Pb-atom during the sliding process: the dimers never switch side.

3. Vertical Manipulation

3.1 XENON ON CU(211)

The procedures involved in vertical manipulation are the following: in order to pick up an adsorbed atom or molecule with the tip, it is again brought close to the adparticle.

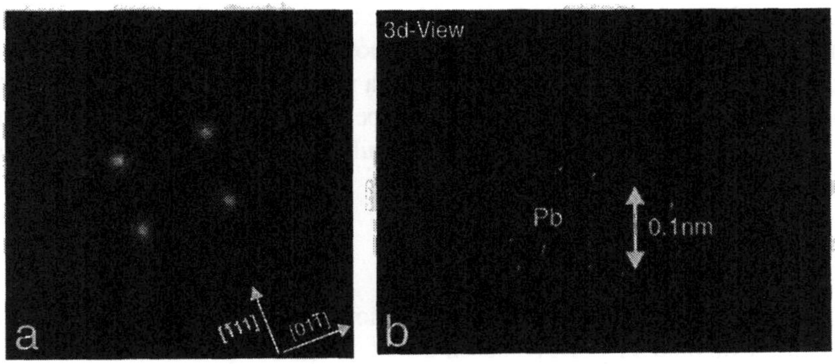

Figure 7. Lateral manipulation of Pb-atoms. The STM images (a: top view, b: 3d-view) show a rectangle formed with four single Pb-atoms. The distance along the rows is six Cu nearest neighbour distances. The height of individual Pb-atoms is 0.1 nm. T = 20 K. Image size: 6 nm * 6 nm.

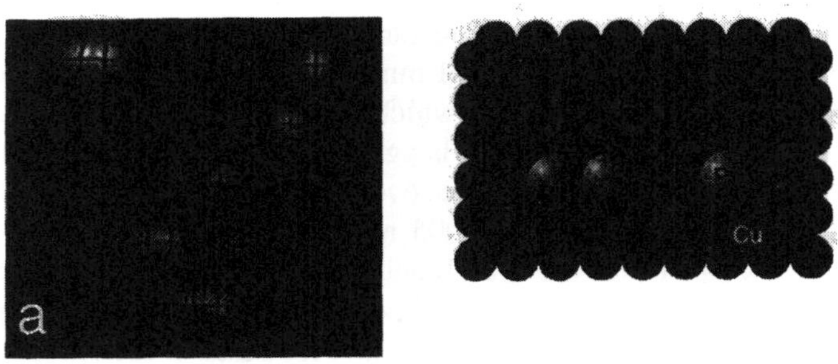

Figure 8. Adsorption sites of Pb-atoms and Pb-dimers determined by registration information. a: The mesh size of the grid corresponds to the size of the unit cell. The center of Pb-dimers is shifted by half a lattice constant along the close packed rows as compared to a single Pb-atom. b: Corresponding sphere model.

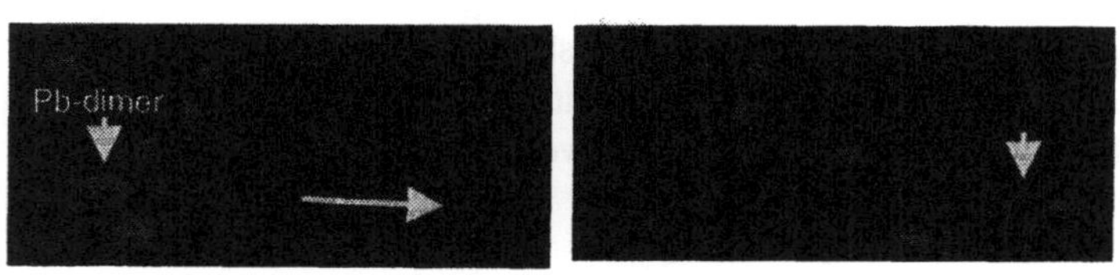

Figure 9. Lateral manipulation of a Pb-dimer together with a single Pb-atom. Note that the Pb-dimer survives the lateral manipulation (for details see text).

The transfer from the surface to the tip is supported by choosing the polarity between substrate and tip in such a way that the electron current is directed from the sample to the tip. According to Eigler et al. [2, 16] heating assisted electromigration is responsible for the transfer to the tip. The fact that the polarity has to be reversed in order to transfer the adparticles back to the surface underlines this conclusion.

Fig. 10 proves that we are also able to perform vertical manipulations routinely with our STM [10]. In this example we transfer a Xe atom to another surface site by carrying it over with the tip at a surface temperature of 15 K. In the topmost STM picture a single Xe atom is clearly visible as the bright spot in the upper right part. In the central picture the Xe atom is not visible as it had been transferred to the tip and the scan is taken with the Xe atom located at the tip apex. In the lowest picture the Xe atom has been put back to the surface at, however, another location. Note, that in the central picture the resolution is much better than in the two others: The step structure of the Cu(211) surface is resolved very well. This is further proof that the Xe atom sits at or (very) close to the tip apex and that it sharpens the tip. By releasing the Xe atom back to the surface we return to the initial shape of the tip, so that the lowest picture is taken with the initial low resolution of the topmost scan. Fig. 11 shows linescans for each image shown in Fig. 10 along the marked black line. Shown is the absolute height with the same absolute background subtraction for all lines. It can be seen that after picking up the Xe atom the tip length has effectively increased by 0.07 nm as the tip withdraws to the imaging position by this extra distance. After removing the Xe atom, the absolute value is again the same within 0.01 nm (this is a measure of the drift). Since the time between the first and the last image is more than 15 min this shows once more the excellent stability of the entire system. It is obvious that with the technique of vertical manipulation we can produce tips of defined geometry and chemical composition at their apex, so that it is possible to study the influence of the tip composition on the imaging process in a systematic way.

3.2. PROPEN ON CU(211)

With less reliability, but using a similar technique as described above, we have been able to reversibly transfer propen molecules between the surface and the tip. Fig. 12a exhibits single C_3H_6 molecules (showing up as the bright elongated hills) and single CO molecules coadsorbed on the Cu(211) surface. A defect step separates the two Cu(211) terraces observed in this image. The single propen molecules are rotated against the close packed rows. Due to symmetry two different orientations are observed. We have not been able to convert one orientation into the other by lateral manipulation. In contrast to ethen we are not able to determine the adsorption geometry, primarily due to the complicated nature of the substrate. The molecule marked with a white ring in Fig. 12a is removed vertically in Fig. 12b. The appearance of the CO molecules has changed (double tip appearance) which indicates that the tip has changed due to the C_3H_6 being located now at the tip. With a reversed voltage pulse (Fig. 12c) the propen molecule can be transferred back to the surface again. It is now located on the upper surface terrace. It is interesting to note that the C_3H_6 molecule is oriented in the same

Figure 10. Example of controlled vertical manipulation: (a) Image with 'blunt' tip. Close packed copper rows are barely resolved. The white point is a Xe atom (marked by the white arrow). (b) The Xe atom has been picked up with the vertical manipulation technique. Imaging resolution has improved, as the Xe atom sharpens the tip apex. (c) The Xe has been put back on the surface and resolution is as in (a) again. Size of each image: 13.5 nm*13.5 nm.

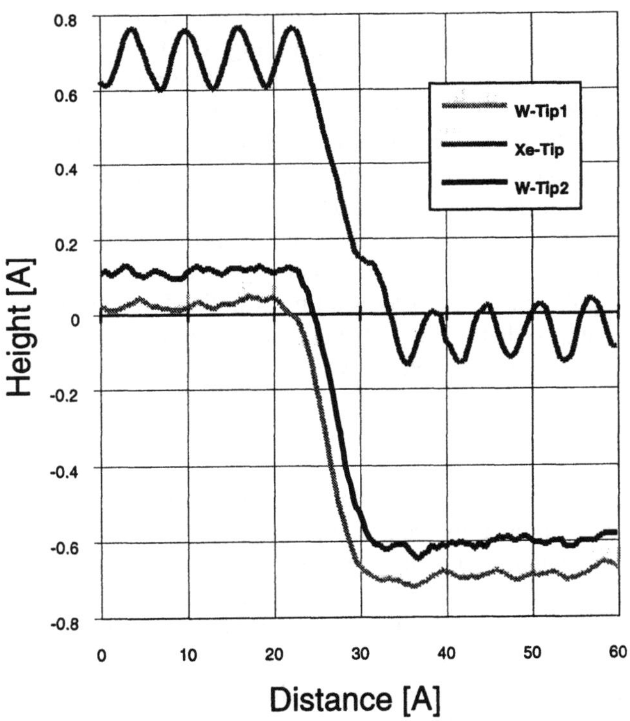

Figure 11. Line scans along the indicated line in the three images of Fig. 10. An increase in tip length of 0.07 nm can be observed after picking up a Xe atom. The small height difference between the scans before and after picking up a Xe atom demonstrates that the vertical drift of the instrument is less han 0.01 nm in 15 min.

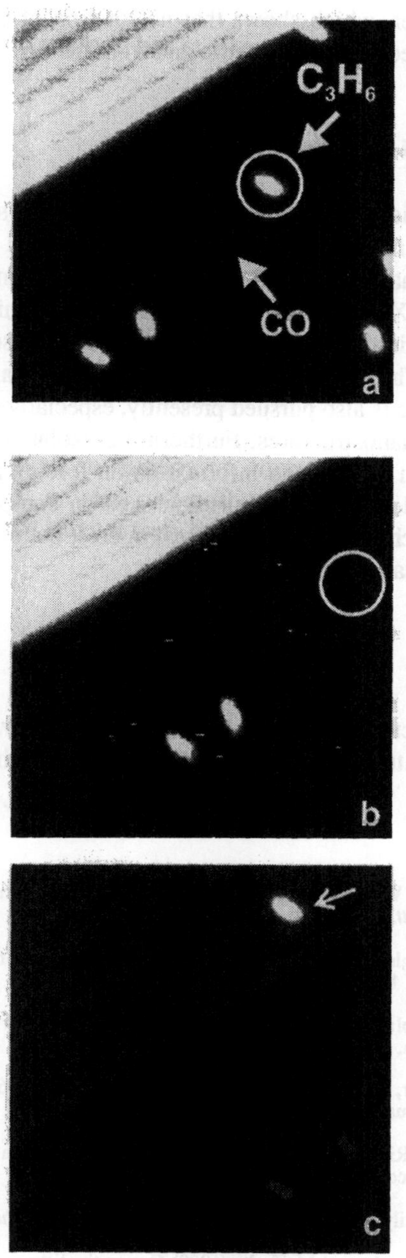

Figure 12. **Vertical manipulation of C_3H_6. The images show the reversible transfer of a propen molecule between the surface and the tip. Image size: 9 nm*9 nm. T =30 K. (For details see text.)**

way as before the transfer, which shows that no rotation of the molecule around the molecular axis has occurred during transfer to and from the tip.

4. Conclusions and outlook

We have demonstrated the manipulation of single atoms and small molecules and the formation of nanostructures at temperatures between 15 and 40 K. As examples for lateral manipulation we have presented CO, C_2H_4, Pb-monomers, and Pb-dimers and for vertical manipulation Xe and C_3H_6. We are currently pursuing coadsorption experiments with different species and the study of diffusion processes by changing the temperature of the sample. The manipulation of single substrate atoms and atom by atom structuring of the substrate is also pursued presently, especially with regard to the build-up of threedimensional nanostructures. Furthermore, isolated substrate atoms can be used as markers to obtain registration information on foreign adsorbates. This will be especially important in determining adsorption sites on complex high index surfaces with large unit cells, which are presently of great interest but not routinely accessible to conventional surface crystallographic methods.

Acknowledgements

We appreciate useful discussions with D.M. Eigler, J. Gimzewski and A. Baratoff. We thank F. Hofmann, G. Witte and J.P. Toennies for lending us their Cu(211) crystal.

References:

1. G. Binnig, Ch. Gerber, E. Weibel, and H. Rohrer (1983) 7 × 7 Reconstruction on Si(111) Resolved in Real Space, *Phys. Rev. Lett.* **50**, 120.

2. J.A. Stroscio and D.M. Eigler (1991) Atomic and Molecular Manipulation with the Scanning Tunneling Microscope, *Science* **254**, 1319, and references therein.

3. P. Avouris (1995) Manipulation of Matter at the Atomic and Molecular Levels, *Acc. Chem. Res.* **28**, 95, and references therein.

4. T.A. Jung, R. R. Schlittler, H. Tang, C. Joachim, J.K. Gimzewski (1996) Controlled Room-Temperature Positioning of Individual Molecules: Molecular Flexure and Motion, *Science* **271**, 181.

5. G. Meyer, B. Neu, K.H. Rieder (1995) Controlled lateral manipulation of single molecules with the scanning tunneling microscope, *Appl. Phys. A* **60** *(Rapid Comm.)*, 343.

6. K. Besocke (1987) An easily operable scanning tunneling microscope, *Surf. Sci.* **181**, 145.

7. J. Frohn, J.F. Wolf, K. Besocke and M. Teske (1989) Coarse tip distance adjustment and positioner for a scanning tunneling microscope, *Rev. Sci. Instrum.* **60**, 1200.

8. S. Ishi, Y. Ohno and B. Viswanathan (1985) An overview of the electronic and vibrational properties of adsorbed CO, *Surf. Sci.* **161**, 349; J.C. Campuzano (1990), in *'The Chemical Physics of Solid Surfaces and Heterogeneous Catalysis'*, Vol. 3, Part A, Eds. D.A. King and D.P. Woodruff, p.389.

9. G. Meyer, B. Neu, K.H. Rieder (1995) Identification of ordered CO structures on Cu(211) using low temperature scanning tunneling microscopy, *Chem. Phys. Lett.* **240**, 379.

10. B. Neu, G. Meyer and K.H. Rieder (1995) Controlled vertical and lateral manipulation of single atoms and molecules with the scanning tunneling microscope, *Mod. Phys. Lett. B* **9**, 963.

11. J. Tersoff and D.R. Hamann (1983) Theory and Application for the Scanning Tunneling Microscope, *Phys. Rev. Lett.* **50**, 1998; N. Lang (1986) Theory of Single Atoms Imaging in the Scanning Tunneling Microscope, *Phys. Rev. Lett.* **56**, 1164.

12. P. Sautet and C. Joachim (1992) Are electronic interference effects important for STM imaging of substrates and adsorbates? - A theoretical anaysis - ; *Ultramicroscopy* **42-44**, 115.

13. O. Schaff, A.P.J. Stampfl, Ph. Hofmann , S. Bao, K.-M. Schindler, A. M. Bradshaw, R. Davis, D.P. Woodruff, V. Fritzsche (1995) Ethene adsorbed on Cu(110): a combined photoemission and photoelectron diffraction study, *Surf. Sci.* **343**, 201.

14. S. Zöphel, G. Meyer and K. H. Rieder (to be published) Manipulation of ethene molecules and lead atoms with the STM.

15. C. Nagl, O. Haller, E. Platzgummer, M. Schmid, P. Varga (1994) Submonolayer growth of Pb on Cu(111): surface alloying and de-alloying, *Surf. Sci.* **321**, 237.

16. D.M. Eigler, C.P. Lutz and W.E. Rudge (1991) An atomic switch realized with the scanning tunneling microscope, *Nature* **352**, 600.

ORGANIC MOLECULAR WIRES: THEORETICAL ANALYSIS OF THE ELECTRONIC COUPLING TO METAL ELECTRODES AND OF THE INFLUENCE OF DERIVATIZATION

J.L. Brédas, V.M. Geskin, D.A. dos Santos, D. Beljonne, and J. Cornil
Centre de Recherche en Electronique et Photonique Moléculaires
Service de Chimie des Matériaux Nouveaux
Université de Mons-Hainaut
Place du Parc 20, B-7000 Mons (Belgium)

I. Introduction

When analyzing the potential usefulness of a given π-conjugated organic compound to act as a molecular wire, there are at least two aspects of major importance from the electronic structure standpoint that need to be considered:

(i) The first is related to the locations of the HOMO and LUMO levels of the molecule with respect to the Fermi level of the metal used as electrode: the energy differences between these levels indeed determine the transmittance through the wire as a function of bias.

(ii) The second deals with the nature of the electronic coupling of the organic molecule to the electrode and the way this can affect the charge injection into the wire.

In this context, a good deal can be learned from the experience that has been gained in the area of polymer-based light-emitting diodes, as well as from the grafting of organic molecules onto metal surfaces: in the former case, the adjustment of the polymer ionization potential (and electron affinity) with respect to the Fermi level of the hole-

C. Joachim and S. Roth (eds.), Atomic and Molecular Wires, 47-59.

injecting (electron-injecting) electrode is essential in order to provide a balanced injection of electrons and holes; in the latter case, one has to understand the parameters that lead to a strong (covalent) bonding of an organic molecule on a metal surface as well as the preferred orientation of the molecule on the surface.

In this contribution, we review two of our recent works that deal with such issues. We first discuss the possibilities of fine tuning the HOMO and LUMO level energies in poly(paraphenylene vinylene), PPV, oligomers, via side-chain derivatization or main-chain substitution. We then describe the grafting process of acrylonitrile on copper. The calculations on the PPV derivatives are performed with a combination of (post) Hartree-Fock type techniques; those on the acrylonitrile-copper system are carried out at the Density Functional Theory (DFT) level, since in the latter case a proper description of the metal electronic structure is required.

II. Influence of derivatization of the conjugated backbone

The achievement of high efficiencies in polymer-based light-emitting diodes, LEDs, requires a proper balance of electron and hole injection rates in order to prevent at best quenching of the excitons by charged defects [1]. One has thus to try and minimize the energy barriers between the Fermi energy of the metallic electrodes and the energy of the frontier levels of the polymer layer; one efficient way to do so is to modulate the positions of the band edges of the polymer through derivatization of its conjugated backbone, for instance with π-electron donor and/or acceptor groups [2-4]. It is interesting to note that the grafting of cyano side-groups on the vinylene moieties along the PPV backbone has led to a significant increase in electron affinity of the polymer, and has therefore favored the use as electron-injecting electrode, of metals with good environmental stability, for instance aluminum. In this context, the fabrication of double-layer devices based on PPV and a cyano-substituted derivative was found to be very efficient since both the minimized energy barriers at the polymer-metal interfaces and the strong confinement of the charge carriers at the polymer-polymer interface contribute to a high quantum efficiency [5]; the best value reported to date stands around 12% [6].

We examine here PPV oligomers containing 5 phenylene rings; the chemical structure of this compound as well as of the derivatized PPV oligomers we have considered are presented in Figure 1. The geometry of the oligomers in their ground state (S_0) has been optimized by means of a semiempirical Hartree-Fock technique derived from the NDDO (Neglect of Differential Diatomic Overlap) formalism, namely the AM1 (Austin Model 1) [7] method; planar conformations are consistently assumed for all the systems. The C-C bond lengths of the PPV oligomers as obtained from AM1 calculations are found to be in very good agreement with X-ray diffraction data [8]. On the basis of the optimized geometries of PPV oligomers and derivatives, transition energies in these systems and related transition intensities are determined with the semiempirical Hartree-Fock INDO (Intermediate Neglect of Differential Overlap) Hamiltonian [9] combined with a single configuration interaction (SCI) technique [10].

Analysis of the one-electron structure provided by the INDO calculations indicates the way the positions of the frontier levels in the five-ring PPV oligomers evolve upon derivatization. Substitution with π-donor groups (derivative **II** in Figure 1) leads to an overall destabilization of the HOMO and LUMO levels, the first being the most affected; in contrast, these levels are stabilized when π-acceptor substituents are attached on the backbone (derivative **III**), the stronger shift being then observed for the LUMO level.

We report in Table 1 the shift in the positions of the HOMO and LUMO levels in the various derivatives reviewed here, where we deal with methoxy and cyano groups as the donor and acceptor entities, respectively. The trends we calculate can be rationalized by a simple three-level model involving the frontier levels of the unsubstituted repeat unit and the occupied (unoccupied) level of the donor (acceptor) group [11]. The system presenting the two kinds of substituent (derivative **IV**) is characterized by an overall stabilization of the frontier levels (that are located in the energy range delimited by the "donor-only" and "acceptor-only" cases); this feature reflects the stronger coupling to the backbone of the

TABLE 1. INDO shifts (in eV) of the HOMO and LUMO levels for derivatives **II** to **VI**, with respect to the unsubstituted PPV pentamer. The energy (in eV) of the lowest optical transition, as calculated at the INDO/SCI level, is reported in the last column

Derivatives	HOMO	LUMO	E_g
PPV			3.24
II	+ 0.27	+ 0.08	3.06
III	- 0.99	- 1.17	3.11
IV	- 0.17	- 0.55	2.92
V	- 0.28	- 0.52	2.92
VI	- 0.32	- 0.19	3.32

cyano group with respect to the methoxy substituents. Cyclic voltammetry measurements indicate that the cyano functionalization on the vinylene moieties lowers the HOMO and LUMO levels by 0.45 eV and 0.62 eV, respectively [12]; these shifts compare very well with the corresponding INDO calculations leading to values of 0.44 and 0.63 eV.

The results thus indicate that an appropriate choice in the nature of the side-groups allows one to modulate the positions of the frontier levels, and hence the color of the emitted light in LED devices [11]. We note that derivatives **V** and **VI** where nitrogen atoms are incorporated within the phenylene rings or the vinylene linkages behave as chains substituted by π-acceptor groups [11].

Calculations conducted at the correlated level provide valuable information on the evolution upon derivatization of the lowest optical transition, and hence of the wavelength of the emitted light. As is the case in the PPV oligomers, the lowest absorption in all the derivatives can be mainly described by an electron transition between the HOMO and LUMO levels; there is, however, one exception in the case of compound **VI**, poly(1,4-phenylenemethylidynenitrilo-1,4-phenylenenitrilomethylidyne), PPI (see Figure 1) for which a mixing of several configurations is needed to properly describe the absorption peak. Note

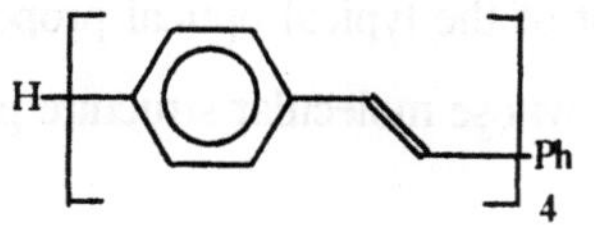

I

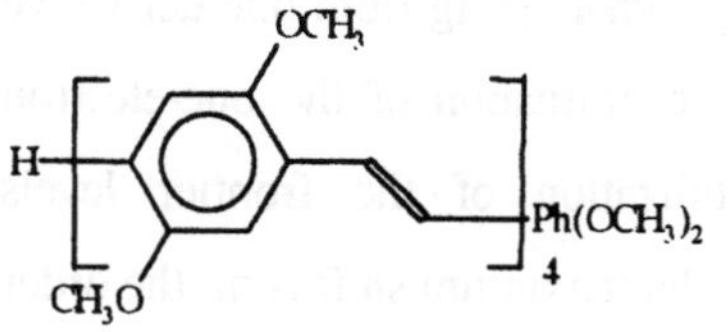

II

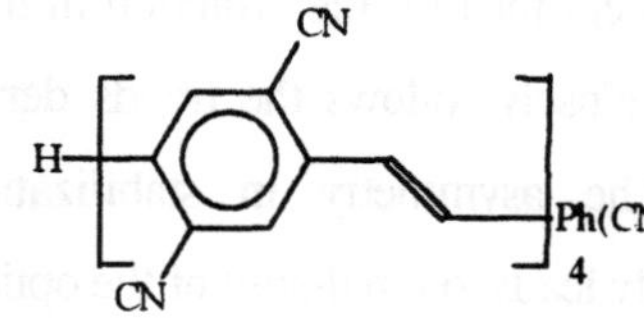

III

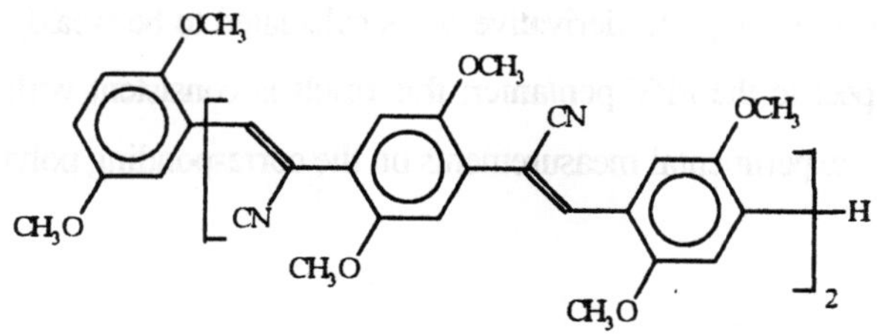

IV

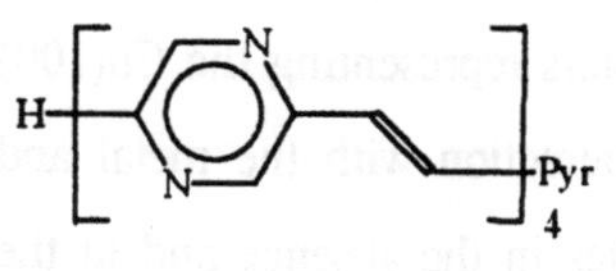

V

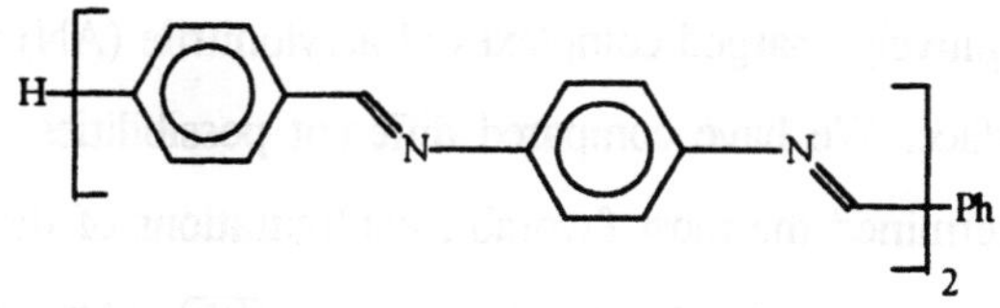

VI

Figure 1. **Molecular structures of the conjugated oligomers reviewed in this work** that such a behavior is reminiscent of the typical optical properties of the emeraldine base form of the polyaniline family [13] whose molecular structure presents similarities with that of derivative **VI**.

Except for PPI, the evolution in transition energy when going from one derivative to the next closely follows the trends derived from the examination of the one-electron structure. The asymmetry in stabilization or destabilization of the frontier levels systematically leads to a red-shift of the optical transition; the maximum shift is on the order of 0.3 eV in the case of derivatives **IV** and **V**, see Table 1. These results rationalize the color modulation observed in LED devices when going from PPV to a derivatized chain as active layer. The lowest transition energy in derivative **VI** is calculated to be weakly blue-shifted, by ~0.08 eV, with respect to the PPV pentamer; this result is consistent with the 0.07 eV blue shift reported from experimental measurements on the corresponding polymer [14].

III. Grafting of acrylonitrile on copper

We now turn to a discussion of the results of a DFT study on neutral and negatively charged complexes of acrylonitrile (AN) with clusters representing the Cu(100) surface. We have compared different possibilities of AN interaction with the metal and determined the most favorable configurations of the adsorbates in the absence and in the presence of an external electric field (EF). Our aim is to evaluate the influence of the electrochemical polarization on the type of AN adsorption and to model the initial stage of electrochemical grafting and polymerization of AN on a copper electrode [15, 16], in relation with the experimental data on the adsorption of AN and electrochemical grafting of polyacrylonitrile.

The copper cluster of choice is the $Cu_9(5,4)$ cluster (where the subscript indicates the total number of copper atoms in the cluster and the numbers in parentheses denote the composition of its layers: here, 5 atoms in the upper layer and 4 in the bottom layer); in this cluster, the upper layer is large enough to study various types of AN adsorption;

furthermore, the central atom has the same number of nearest neighbors as on the actual (100) surface. Therefore, this cluster will be our system of choice. However, since it will be shown that the AN molecule can be bound to two metal atoms, we also consider the $Cu_{14}(8,6)$ cluster, in which both central copper atoms have the right coordination number.

The non-polarized electrode is simply modeled by not considering any external EF (we also neglect the effect of the electrolytic medium). On the contrary, an EF is taken into account to mimic the electrochemical polarization; the direction of EF is set normal to the (100) cluster plane and defined as the z-direction with the metal taken as the negative pole (cathode). The aim of introducing EF is to cause a reasonable charge transfer from the cluster to an AN monomer; therefore, the criterion chosen for the value of EF is that the charge on AN be about -1. We note that our preliminary tests have shown that for a given complex: (i) the charge on the AN molecule is virtually independent of its geometry (in particular, it does not change during geometry optimization from a reasonable starting geometry); (ii) the charge transfer evolves approximately linearly with EF (which facilitates the setting of the EF value); and (iii) the charge transfer in a given EF does slightly depend on the cluster size, a one-electron transfer requiring a field in the range $0.8\text{-}1.0 \times 10^8$ V/cm (these are very reasonable field values when comparing to those estimated in nonlinear optical measurements in solutions [17]).

The calculations were performed in the framework of the DFT method, using the DMol program [18], with a double zeta numeric basis set including polarization functions. Geometry optimizations are carried out within the local spin density approximation (LSD) using the Vosko-Wilk-Nusair functional (denoted VWN); the geometry optimizations were unconstrained except for the distances between copper atoms that were kept at the bulk crystal values (nearest-neighbor distance of 2.551 Å). Although the overall description of binding within LSD calculations is good, this approach shows a strong tendency to overestimate binding energies; therefore, the binding energies reported here are calculated including non-local corrections in two ways. In the first one, denoted BVWN, Becke's 1988 version of a gradient-corrected exchange potential corrections are introduced in a final non-self-consistent evaluation of the energy at a negligible additional computational cost, while only the VWN potential is used throughout the SCF iterations. In addition, the binding energies of the key structures are calculated in the most precise and computationally demanding way using throughout the SCF iterations both the gradient-corrected exchange potential by Becke and the gradient-corrected correlation by Perdew and Wang; this approach is hereafter noted BP. The results for the representative AN-copper complexes are collected in the Table 2.

In the absence of EF, the geometry of adsorption is very similar for the neutral complex and its negative counterpart since the charge on AN does not differ significantly in both cases; the geometry does not depend much on the cluster size as well. However, the energy of adsorption becomes less favorable as the cluster size and the coordination number of the metal atoms directly bound to the organic molecule increase: the more saturated the copper atoms in the cluster, the lower their tendency to form new chemical bonds. The optimized structure is a complex with the C=C and CN groups asymmetrically coordinated each at one copper atom (as an example, AN-Cu_9(5,4) is shown in Fig. 2a). Only two copper atoms show distances with AN atoms indicating the formation of chemical bonds; the shortest contacts are with the two terminal backbone atoms. This complex is rather a di-σ-complex than a di-π-complex of Dewar's type. As a matter of fact, experimental data [19, 20] show that on a group I transition metal, AN is most likely adsorbed with both vinyl and nitrile groups coordinated at metal atoms. The changes in bond lengths in AN, relative to the isolated molecule, are actually similar to those calculated in the AN radical-anion: C^1C^2 and C^3N strongly elongate, while C^2C^3 assumes a double-bond character.

Upon application of EF, qualitatively, two opposite factors determine the orientation and the geometry of the organic molecule in the complex with the metal cluster. The metal tends to bind to several backbone atoms, as we have seen in the absence of EF, and thus to align the molecule along the cluster surface. At the same time, the field tends at first to align the dipole of the molecule along the field, *i.e.* normal to the cluster surface, which requires breaking some of the bonds with the metal; finally, it tends to pull the AN anion away since its total energy decreases in the direction opposite to the field, thus dissociating the complex. The items relevant for the evaluation of the grafting mechanism

TABLE 2. Major characteristics of several acrylonitrile-copper complexes. The bond lengths are in Å, the Hirshfeld charge on the AN moiety (q) in |e| units, and the binding energies (E_b) in kcal/mol. The binding energy, in case of negative complexes, is evaluated relative to the energies of the negative cluster and neutral AN (which is the most favorable combination of isolated components) and, in the presence of electric field, for the z=0 position of the upper copper layer; a negative value implies stabilization of the complex with respect to its components.

	Bond lengths							q	E_b	
	C^1C^2	C^2C^3	C^3N	C^1Cu	C^2Cu	C^3Cu	NCu		BVWN	BP
[AN-Cu_{14}] on edge	1.437	1.362	1.202	2.116	2.619	2.473	1.937	-0.34	36.9	15.3
[AN-Cu_9] on edge	1.427	1.362	1.201	2.075	2.847	2.375	1.893	-0.34	29.2	1.5
[AN-Cu_9]$^-$ on edge	1.432	1.361	1.208	2.074	2.490	2.377	1.898	-0.49	28.1	1.6
[AN-Cu_9]$^-$ on top F_z=0.77×10^8 V/cm	1.429	1.375	1.200	2.179	3.165	–	–	-0.89	-2.4	-31.6
[AN-Cu_9]$^-$ on edge F_z=0.77×10^8 V/cm	1.473	1.358	1.210	2.051	2.626	2.432	1.961	-0.85	27.9	-15.1
[AN-Cu_9]$^-$ on edge F_z=1.03×10^8 V/cm	1.488	1.360	1.212	2.086	2.738	2.485	1.939	-1.01	23.9	

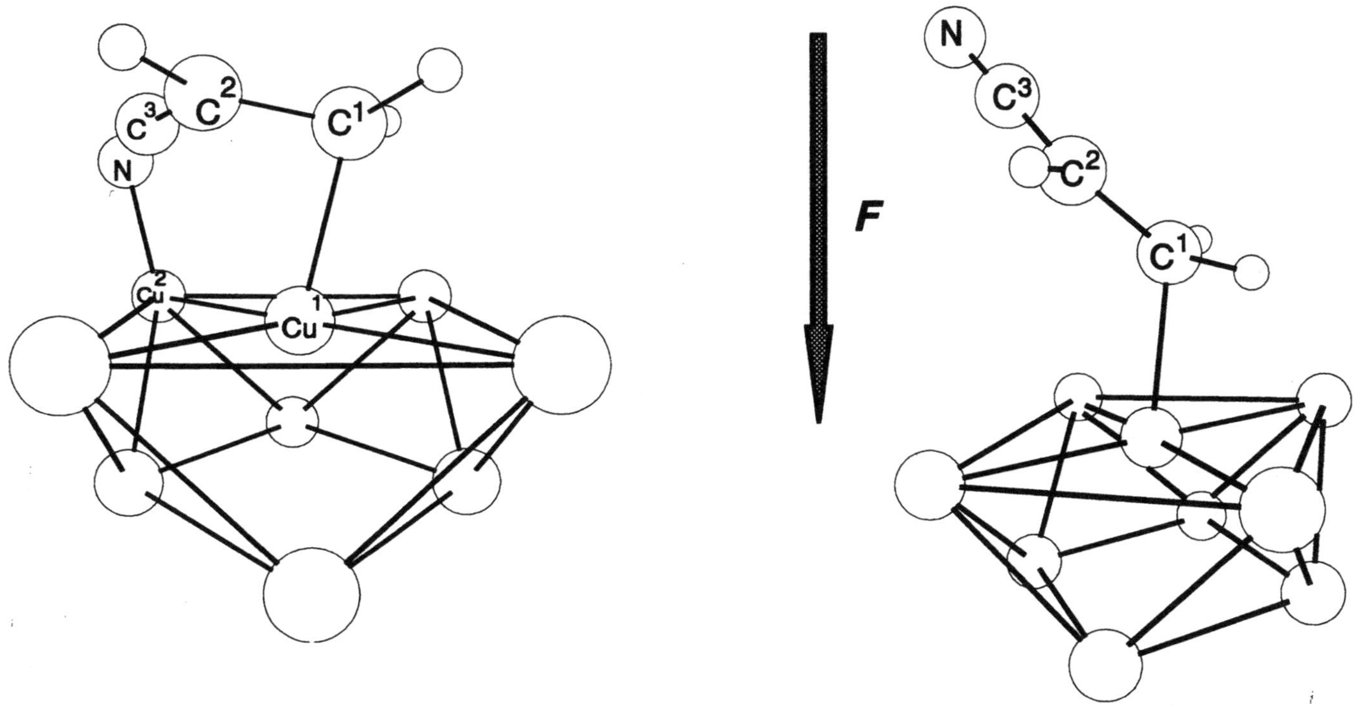

Figure 2. Acrylonitrile - copper cluster complexes: (a) AN-$Cu_9(5,4)$ in the absence of electric field; (b) $[\text{AN-}Cu_9(5,4)]^-$ in the presence of an electric field of 0.77×10^7 V/cm

are, therefore, to estimate whether: (i) there is a bound state for AN and AN^- on copper clusters in a given EF; (ii) there is a (local) minimum corresponding to a σ-complex; (iii) multiply-coordinated complexes survive in EFs; (iv) there are multiple minima.

With $Cu_9(5,4)$ in a field of 7.7×10^{7} V/cm, we find that the charge carried by the AN moiety in negatively charged complexes is on the order of -0.9 |e|, *i.e.* the charge transfer is almost complete. Two potential minima are found [21].

The first one is a complex "on-edge" corresponding to that obtained in the absence of EF (shown in Fig. 2a), with C^1 bound to one copper atom and N to another. In comparison to the zero-field complex, the bond to N is slightly longer, C^3 and C^2 are also slightly further away from copper, reflecting the tendency of the nitrile group to align with the field. Within the AN moiety, the C^1-C^2 bond elongates, while the C^2-C^3 bond length does not change. C^2 is not bonded to copper and remains rather unsaturated; it could therefore be the active site for the initiation of AN polymerization.

The second and more stable complex (Fig. 2b) has no counterpart in the absence of EF. It has only one bond with the cluster, namely, C^1 is bound to one copper atom, while the rest of the molecule tends to align along EF: it can be referred to as a σ-complex. All the bond lengths within the AN moiety are close to their values in isolated AN^-, except for C^1-C^2 that is slightly longer, which is due to the bonding of C^1 with Cu. Therefore, the reactivity of C^2 in this complex can be at least the same or slightly higher than in non-coordinated AN^-. The bonds around C^2 are practically coplanar, thus indicating sp^2 hybridization, while C^1 shows considerable deviation towards sp^3 hybridization. A certain π-bonding along C^1-C^2 still persists. What prevents CN against aligning is not the steric interaction between H^3 and copper since the nearest H-Cu distance is about 3.4 Å in the optimal geometry and 2.8 Å if aligned. We are therefore led to the conclusion that the rotation around C^1-C^2 is still significantly hindered, and this effect is still larger than the aligning action of EF. Since the second complex is more stable than the first one (by ca. 17 kcal/mol at the BP level), we conclude that the gain in energy due to dipole orientation and shift in effective charge center upon EF are energetically more important than the dissociation of the Cu-N bond. The anticipated rearrangement towards this complex corresponds to the field-induced reorientation and grafting postulated previously [15]. However, we believe that the complex "on-edge", as well as this one, could be an intermediate for further polymerization of AN.

When increasing the EF value to 10^8 V/cm, the charge on AN becomes slightly higher than -1 |e|, *i.e.* the charge transfer is full. In such a field, the doubly coordinated complex does survive; its geometry changes further along the same direction as discussed above for the on-edge complex. However, there is no longer any bound state for the σ-complex: the AN radical ion rather tends to leave the cluster. The fact that it would have only one loose bond with the cluster, together with the alignment of its dipole, appears to ease the dissociation. We believe this result is important and can be directly related to the actual experimental situation. First, as we have shown, this type of complex is the most favorable at low EF; therefore, it is likely to be the initial state when the field is increased. Second, this way of desorption is intuitively realistic. Finally, it coincides well with the experimental observation of the absence of polymer grafting when the reaction is carried out at a very high cathodic potential, corresponding to a higher relative EF.

It is clear that the methodological experience we have gained in analyzing the interaction between a conjugated organic system such as acrylonitrile, and a metal electrode, here copper, proves useful when examining the coupling of organic molecular wires onto metals. On that basis, we are currently investigating for instance the interaction between oligothiophenes and aluminium.

Acknowledgements. This work is partly supported by the Belgian Government "Service des Affaires Scientifiques, Techniques et Culturelles" (SSTC) in the framework of "Pôle d'Attraction Interuniversitaire en Chimie Supramoléculaire et Catalyse", the European Commission ESPRIT project LEDFOS, FNRS-FRFC, and an IBM Academic Joint Study. DB is Chargé de Recherches and JC Aspirant of the FNRS.

REFERENCES

[1] L.S. Swanson, J. Shinar, A.R. Brown, D.D.C. Bradley, R.H. Friend, P.L. Burn, A. Kraft and A.B. Holmes, Phys. Rev. B, **46**, 15072 (1992).

[2] D. Braun and A.J. Heeger, Thin Solid Films, **216**, 96 (1992).

[3] S. Doi, M. Kuwabara, T. Noguchi and T. Ohnishi, Synth. Met., **55-57**, 4174 (1993).

[4] J.L. Brédas and A.J. Heeger, Chem. Phys. Lett., **217**, 507 (1994).

[5] N.C. Greenham, S.C. Moratti, D.D.C. Bradley, R.H. Friend and A.B. Holmes, Nature, **365**, 628 (1993).

[6] D.R. Baigent, N.C. Greenham, J. Grüner, R.N. Marks, R.H. Friend, S.C. Moratti and A.B. Holmes, Synth. Met., **67**, 3 (1994).

[7] M.J.S. Dewar, E.G. Zoebisch, E.F. Healy and J.J.P. Stewart, J. Am. Chem. Soc., **107**, 3902 (1985).

[8] C.J. Finder, M.G. Newton and N.L. Allinger, Acta Crystallogr., Sect. B, **30**, 411 (1974).

[9] J.A. Pople, D.L. Beveridge and P.A. Dobosh, J. Chem. Phys., **47**, 2026 (1967).

[10] J. Ridley and M.C. Zerner, Theor. Chim. Acta, **32**, 111 (1973).

[11] J. Cornil, D.A. dos Santos, D. Beljonne and J.L. Brédas, J. Phys. Chem., **99**, 5604 (1995).

[12] S.C. Moratti, D.D.C. Bradley, R. Cervini, R.H. Friend, N.C. Greenham and A.B. Holmes, SPIE Int. Soc. Opt. Eng., **2144**, 108 (1994).

[13] B. Sjögren and S. Stafström, J. Chem. Phys., **88**, 3840 (1988).

[14] C.J. Yang and S.A. Jenekhe, Chem. Mater., **3**, 878 (1991).

[15] G. Lécayon, Y. Bouizem, C. LeGressus, C. Reynaud, C. Boiziau, and C. Juret, Chem. Phys. Lett. **91**, 506 (1982).

[16] R. Jérôme, M. Mertens, and L. Martinot, Adv. Mater. **7**, 807 (1995).

[17] S.R. Marder, C.B. Gorman, F. Meyers, J.W. Perry, G. Bourhill, J.L. Bredas, and B.M. Pierce, Science **265**, 632 (1994).

[18] D. Mol, version 2.3.5. SanDiego: Biosym Technologies, 1993.

[19] G. Xue, J. Dong, J. Zhang, and Y. Sun, Polymer **35**, 723 (1994).

[20] P. Parent, C. Laffon, G. Tourillon, and A. Cassuto, J. Phys. Chem. **99**, 5058 (1995).

[21] V. M. Geskin, R. Lazzaroni, M. Mertens, R. Jérôme, and J. L. Brédas, J. Chem. Phys., in press.

SYNTHESIS FOR MOLECULAR ELECTRONICS: RIGID RODS AND DISCS

K. MÜLLEN*, S. VALIYAVEETTIL*, V. FRANKE, V. S. IYER
Max-Planck Institute for Polymer Research
Ackermannweg 10
D-55128 Mainz
Germany

Key Words: Molecular Electronics, Wires and Discs, Conjugated Polymers, Graphite Subunits, Poly(*para*-phenylene-ethynylene), 5-Alkoxyisophthalic acid, Scanning Tunnelling Microscopy, Self-assembly, Hydrogen Bonds.

Abstract: Recent developments in molecular electronics warrant new molecular building blocks to deal with problems such as efficient transport charge and input-output processes. This paper deals with our synthetic approaches toward new molecules which involve dithiol functionalised poly(*para*-phenylene-ethynylene)s (PPEs) and processible graphite subunits. The last section is concerned with the self-assembly of molecules on a solid substrate using scanning tunnelling microscopy. The effect of light on the self-assembly of buiding blocks containing azobenzene groups is also discussed.

C. Joachim and S. Roth (eds.), Atomic and Molecular Wires, 61-79.

1. Introduction

The synthesis of nanostructures has attracted considerable attention due to their projected applications[1,2], one of which involves molecular electronics, where mobility of electrons through these structures plays a significant role[3]. In order to facilitate efficient flow of electrons, a conjugated medium is necessary. Since the presence of trace impurities and structural defects cause major problems, synthetically pure molecular components are of great interest in this area due to their high component density, increased response speeds and high energy efficiency[4]. Moreover, with the possibility of increased data storage within molecular components, a large reduction in size of molecular devices compared to the existing counterparts, can be expected[5]. In order to achieve useful organic based molecular electronics, a few parameters need to be addressed such as, a) transmission and storage of the signals (i.e. read and write) in molecular components and b) connection of the molecular components to an external circuit. Our interest in this area is to design new molecular and macromolecular building blocks and to deal with some of the problems of electron transport and the input-output junctions.

Electron or hole transport can be along a conjugated backbone of a polymer or through a column of π-stacked molecules. In this report, we discuss a modular synthesis of molecules having the shape of wires and discs, and their ability to self-assemble in a controlled manner. Furthermore, we use intermolecular interactions such as strong electrostatic and weak van der Waals forces to achieve the desired supramolecular lattice structure.

2. Results and Discussion

2. 1. SYNTHESIS OF MOLECULAR WIRES

Polymers with a conjugated backbone have previously been referred to as molecular wires[6]. A molecular wire can be defined as a conjugated one-dimensional molecule for electron or hole transport between its two ends.

Recently, the design and synthesis of conjugated oligomers/polymers with well defined structure and shape has attracted attention[7]. Since the conjugated system involves a series of completely overlapped π-orbitals, efficient electron delocalisation through this medium is possible. Most of the conjugated polymers have polyparaphenylene (PPP) (**1**) backbone which involves step-ladder (**2**), ladder-type (**3**) and copolymers of PPP with a ladder-type polymer (**4**) (Scheme 1). Soluble PPPs, **1**, are particularly interesting because of their promising optical and electronic properties [8]. High molecular weight PPPs are synthesised by Suzuki-type coupling of bifunctional dibromo- and diboronic acid monomers [9]. From monomers substituted with alkyl chains, highly soluble rigid PPPs are obtained [10]. These substituted PPPs form well organised domains both in solution and in the melt [11].

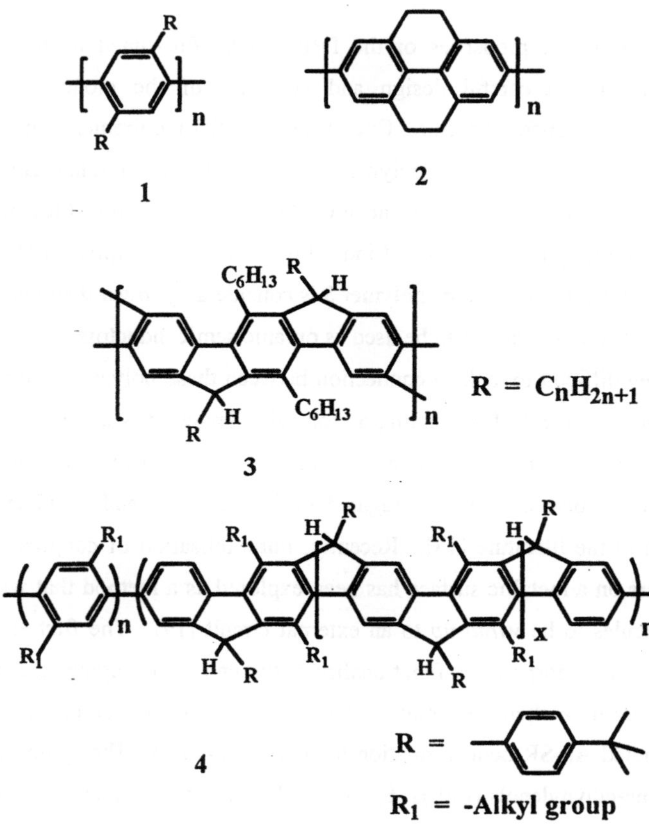

Scheme 1

The persistence in shape of PPP both in the solid lattice and in solution can be increased by introduction of a bridge between the adjacent phenylene rings along the chain. This gives rise to a double stranded topology for the PPP. Polymer **2**, synthesised recently in our laboratory, is considered as a step-ladder polymer[12]. Synthesis of ladder-type PPPs was recently reported by Scherf and Müllen [13]. Our strategy enabled us to form a defect-free structure via a two-step synthesis. First, a PPP with functional groups, that could undergo polymer-analogous cyclizations was synthesised and after the quantitative cyclization step, a high molecular weight ladder-type polymer **3** was obtained. The X-ray analyses of these polymers indicate well organised polymer chains in the solid lattice [14]. They show an intense blue photoluminescence in solution and in addition a yellow electroluminescence, which is characteristic of an aggregate emission. This yellow emission was suppressed in those ladder polymers that contained a spacer group between the planar segments. The copolymer **4** gives only a blue emission with a λ_{max} = 450 nm [15].

Even though the electronic properties of the PPP can be fine-tuned using the ladder-type architecture, a careful design and synthesis of the monomer is necessary to obtain a defect-free structure. One of the essential requirements in the step-wise synthesis of a ladder-type polymer is the choice of a quantitative polymer-analogous cyclisation reaction to achieve the desired structure. Previous results from our laboratory have shown that indeed this can be accomplished [16]. As a result, the band gap of ladder-type polymer **3** is considerably lower than that of the PPP **1** [17]. These polymers can be used as organic semiconductors.

However, it is often difficult to make a connection between these polymers and an external device. In the context of fabricating a molecular device, it is necessary to connect both ends of these molecules to an external *circuit* to achieve an efficient input-output process. Polymers with conjugated backbone and metal-based end groups are known in the literature [18]. Recently, immobilisation of conjugated polymers/oligomers on a metallic surface has been explored as a method that will enable these molecules to be *wired up* to an external circuit [19]. The first step towards this goal is to incorporate a functionality at the end of a polymer chain, that subsequently forms a chemical bond with the surface. In our approach, we chose the well studied Au-SR bond formation to achieve this goal. The synthesis of poly(p-phenylene-ethynylene)s (PPE)s **5**, with -SH functionality as end groups is described next (Scheme 2) [20].

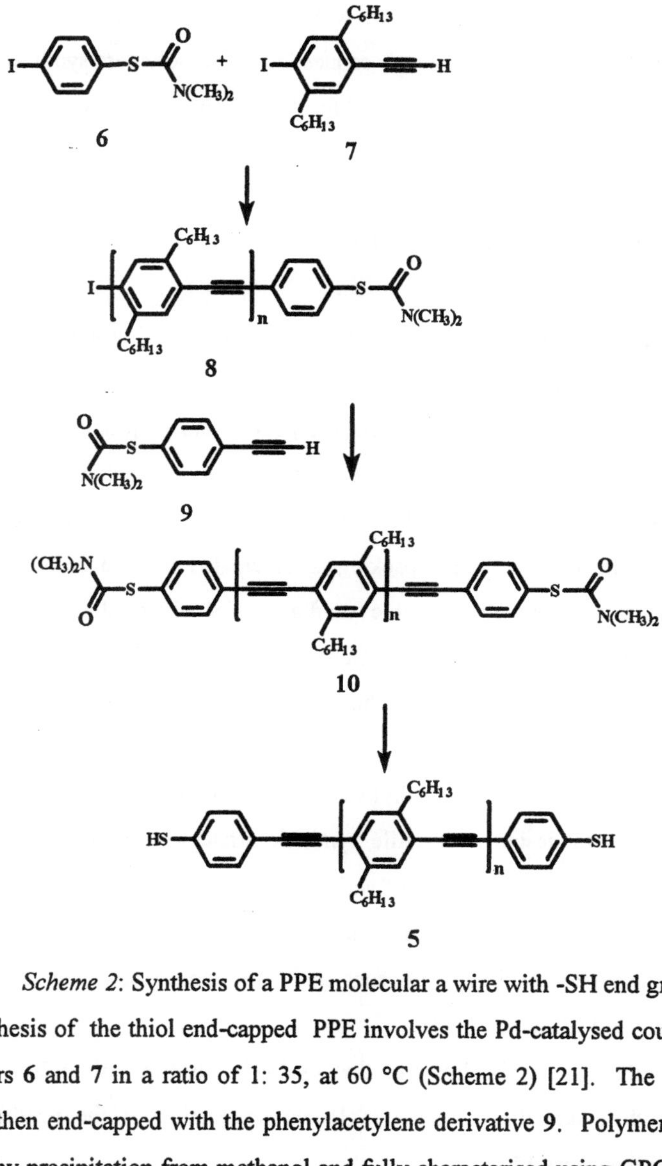

Scheme 2: Synthesis of a PPE molecular a wire with -SH end groups

The synthesis of the thiol end-capped PPE involves the Pd-catalysed coupling of monomers **6** and **7** in a ratio of 1: 35, at 60 °C (Scheme 2) [21]. The polymer chain is then end-capped with the phenylacetylene derivative **9**. Polymer **10** was purified by precipitation from methanol and fully characterised using GPC, NMR, IR, and UV-Vis spectroscopy. Hydrolysis of the thioamide functionality gave the thiol end group, which is necessary for immobilisation of polymer **5** on to gold surface. The polymers synthesised have a molecular weight (M_n) of 12,000 and a polydispersity of 3. It is envisaged that polymers with a narrow molecular weight range are more useful than those with broad distribution in order to connect two gold electrodes. However, if we use gold electrodes with uneven surface morphology, it is conceivable that polymers with broad distribution offer a better chance to obtain the maximum number of connections. The polymers synthesised in our laboratory are currently being examined for their use as molecular wires.

2. 2. SYNTHESIS OF MOLECULAR DISCS

In the second approach, we discuss the synthesis of well defined molecules with disc-type architecture, which unlike the previous one-dimensional systems, offer the prospect of two-dimensional conductors. The goal of the project is to obtain processible and soluble graphite subunits that can be characterised using spectroscopic methods. On account of their large size and possible π-π interaction between the adjacent discs, columnar packing is anticipated from these building blocks. These materials are also suitable candidates for photoinduced charge transport along the stacking axis [22]. Currently, there are several ways to synthesise various disc-type molecular building blocks; some of them are discussed below.

Recently, we reported the synthesis of alkyl substituted graphite subunits [23]. Our goal in this area is to design a versatile synthetic strategy that allows us to synthesise processible graphite subunits of various dimensions in high yield and to explore their properties. As a result of our study, highly soluble, and thereby fully characterisable hexa-peri-hexabenzocoronenes and their higher homologues are available in high yield (Scheme 3) [23].

Key steps in this synthesis involve the cyclotrimerisation of 1,2-bis(4-alkylphenyl)acetylene and the oxidative cyclodehydrogenation of the resulting hexa(alkylphenyl)benzenes **12**.

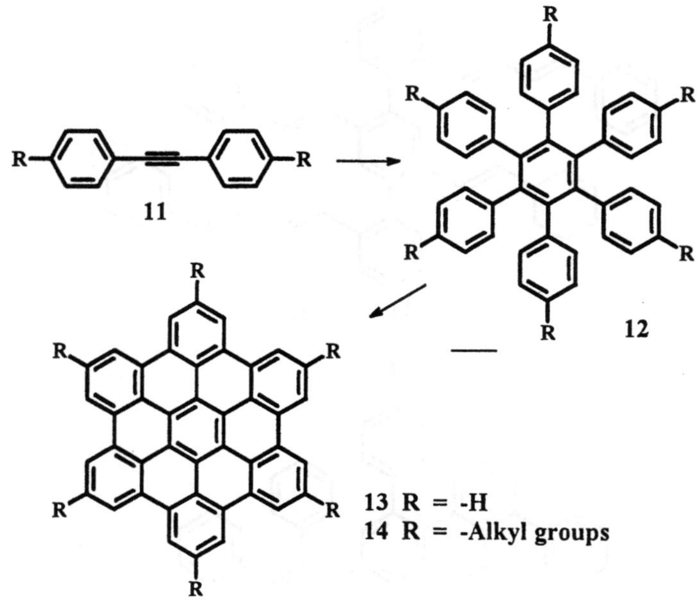

Scheme 3: Synthesis of hexalkylhexabenzocoronene

These compounds, **12** and **14**, were characterised by NMR, Mass and UV-Vis spectroscopy. Redox properties of compound **14** were studied using cyclic voltammetry. Compound **14** also forms a thermotropic mesophase. Differential scanning calorimetry (DSC) analysis revealed a melting point of 45°C and a clearing temperature of > 400°C. X-ray diffraction studies indicate a columnar mesophase with hexagonal superstructure [24]. The existence of such a broad temperature range, highly stable columnar mesophase can be exploited to increase charge carrier mobility and photoconductivity [22].

In order to obtain even larger planar polyaromatic structures, two other synthetic routes were explored. The first involves a cycloaddition-cyclodehydrogenation sequence (Scheme 4) [25].

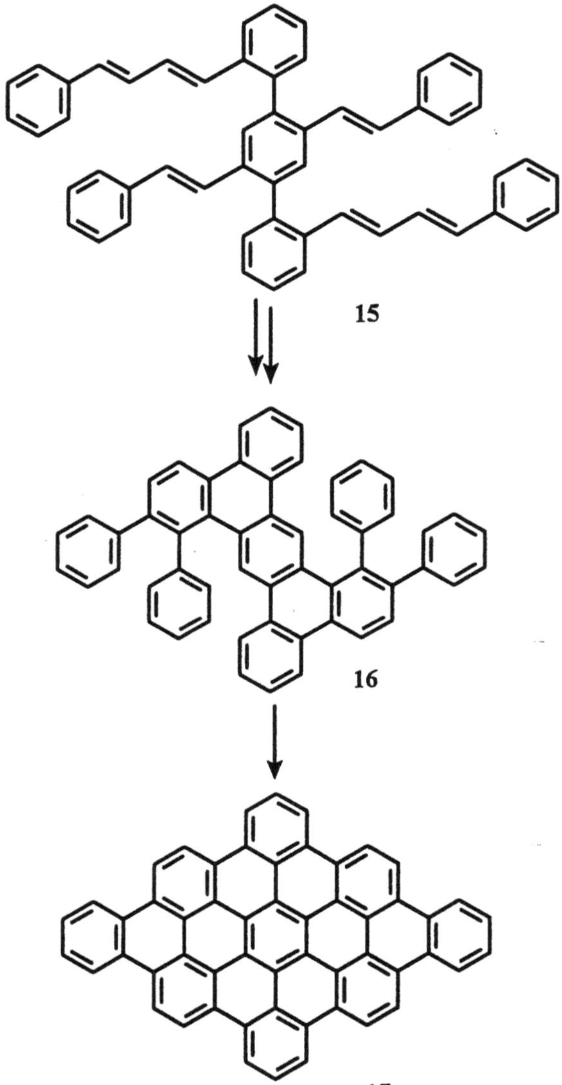

17

Scheme 4: Synthesis of extended graphite subunits

The precursor **16** was characterised using a variety of spectroscopic techniques. Purification of the final product **17** was achieved by high vacuum sublimation. A bright orange microcrystalline product was obtained with high purity that showed an orange red fluorescence upon irradiation with UV light.

The other more versatile route for soluble graphite subunits includes Diels-Alder reactions between substituted cyclopentadienones and acetylene derivatives (Scheme 6). Some of the many interesting complex shapes obtainable through this simple synthetic strategy using readily available starting materials, are shown below (Scheme 5) [26].

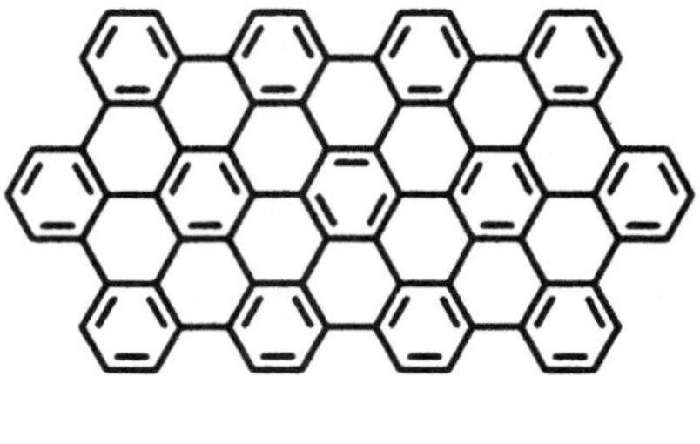

18

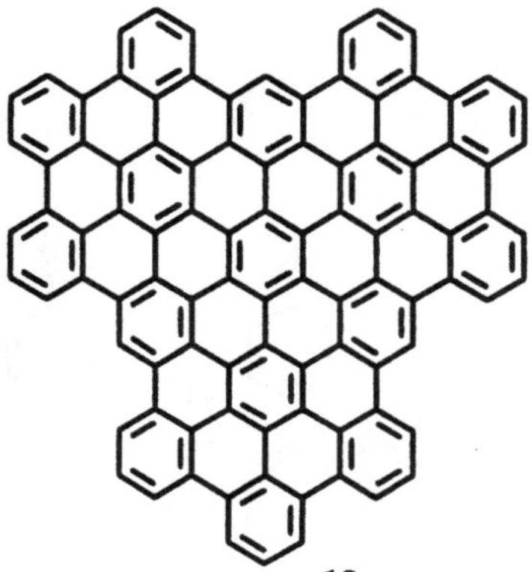

19

Scheme: 5 Molecular structures of graphite subunits

Here also the final step involves a cyclodehydrogenation process in which multiple rings are formed from a soluble precursor. The advantage of using cyclodehydrogenation as a multiple fusion process in the final step is that it allows us to characterise the highly soluble precursor with conventional techniques. Another merit of this route is the flexibility in functionalising the periphery of these large molecules. This can be easily achieved by using properly functionalised phenyl derivatives of the cyclopentadienones **23** or of the acetylene derivative **22**. The characterisation and processing of these polymers depend on their solubility and melting behaviour. This strategy can be further utilised to synthesise several graphite subunits, which differ in size, morphology and peripheral functional groups [26]. The above mentioned building blocks with a wire or disc-type architectures are thus available by new synthetic methods and careful choice of starting materials.

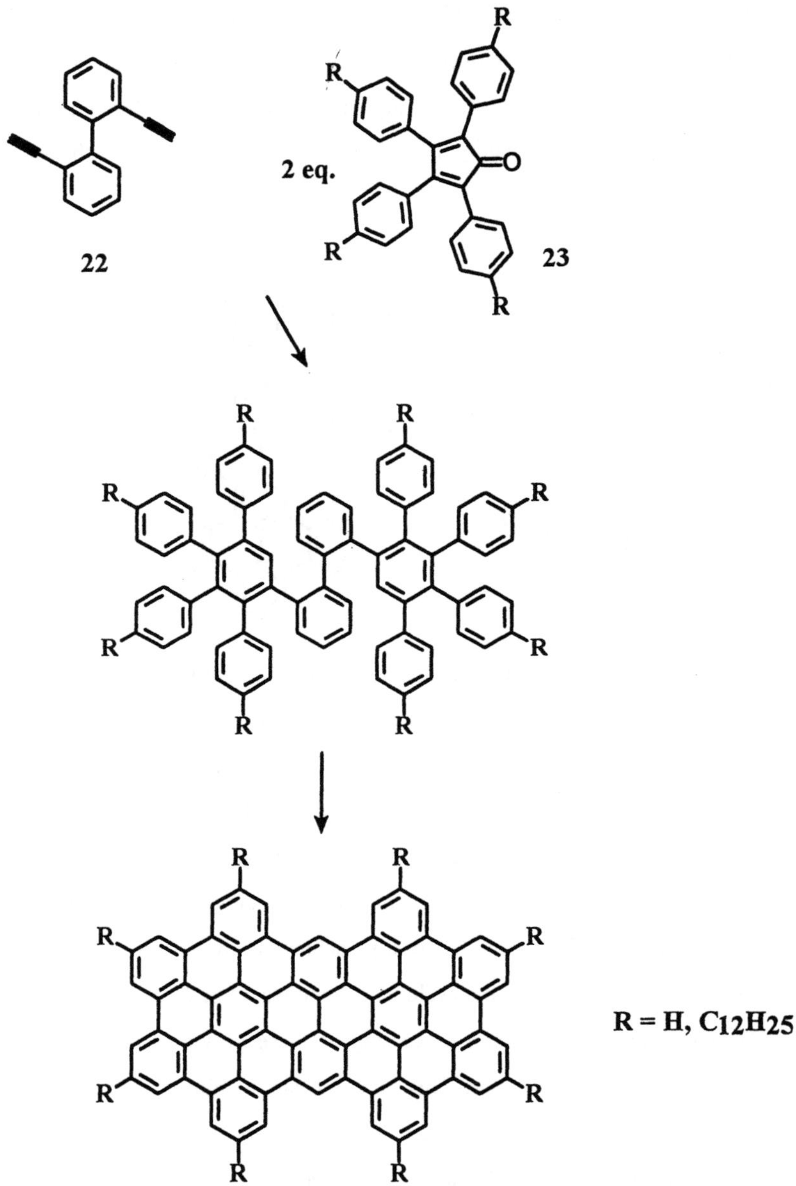

Scheme 6: Synthesis of a "all carbon" macromolecule

2. 3. SUPRAMOLECULAR ARCHITECTURES VIA SELF-ASSEMBLY APPROACH

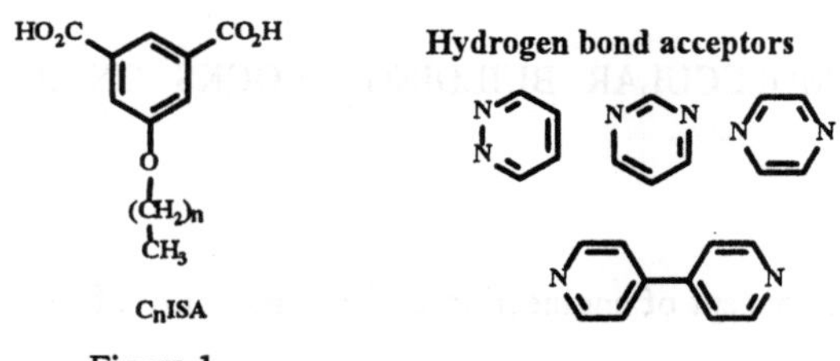

Figure 1

In order to establish a more simple route towards a well-defined morphology, self-assembly of small molecules with complementary functional groups has been explored [27]. Here the molecules are connected through relatively weak non-covalent hydrogen bonds that can be formed reversibly. These bonds are weak enough to be rearranged for correction of an error made during the self-assembly process [28]. Recently, we reported the formation of layers, ribbons and channel-type architectures in the crystal lattice obtained from the self-assembly of simple building blocks such as 5-alkoxyisophthalic acid (CnISA, Fig. 1) and its complexes with bifunctional hydrogen bond acceptors [29].

Here, both hydrogen bonding and alkyl chain crystallisation influence structure formation [29]. For example, CnISA with $6 \geq n \geq 12$, show a layer-type morphology (Fig 2, left), whereas the molecules with $6 \leq n \leq 12$ give rise to a cyclic structure that upon stacking forms a columnar lattice (Fig. 2, right) [29].

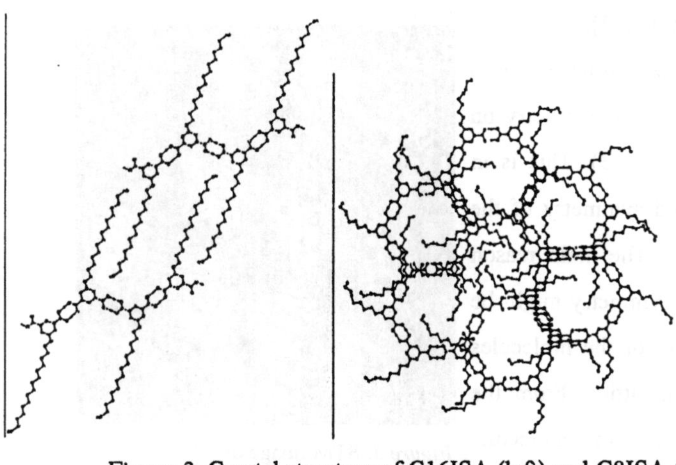

Figure 2: Crystal structure of C16ISA (left) and C8ISA (right)

These structures can also be created in other phases including mesophase and in solution, which is necessary in establishing the structure-property dependence of bulk materials [30].

2. 4. SELF-ASSEMBLY OF MOLECULAR BUILDING BLOCKS ON A SUBSTRATE

It is also important to address the concept of engineering the self-assembly of the above mentioned molecular building blocks at the liquid-solid interface. This type of self-assembly in epitaxial synthesis of molecular thin films may find wide application in sensors and in both conducting and insulating coatings for sensitive components of various instruments and in molecular electronics [31]. Recent developments in scanning tunnelling microscopy (STM) allow us to study these monolayers with submolecular resolution and to perform manipulations [32]. It is also proposed that an STM tip would be able to separate, remove, add, and/or reposition atoms, molecule or ions. These capabilities may allow us to store and manipulate data in nanoscale dimensions and to construct circuits, that are crucial for the development of molecular electronics [3f].

Previously, a variety of techniques have been used to create monomolecular layers of the above mentioned building blocks on substrates, most predominantly adsorption from solution on to graphite substrate and vacuum deposition on molybdenum disulphide [33].

Hexadodecylhexabenzocoronene gives a monolayer with a twofold symmetry on a graphite substrate (Fig. 3). This is in contrast to the sixfold symmetry of the molecule itself [23]. The main reason for this reduction in symmetry might be the packing constraint of the molecules in a two-dimensional lattice. From the current-potential curve of this molecule, an unambiguous distinction between the

Figure 3: STM image of hexadodecylhexabenzocoronene

aliphatic and aromatic regions on the graphite can be made; in addition, the curve taken over the aromatic domain is asymmetric by one order of magnitude and possesses the shape of a diode function [23].

One possible explanation may be that the highest occupied molecular orbital of hexabenzocoronene is energetically close to the Fermi-level of the electrode, and thereby a resonant amplification of the current is observed [23]. It is also possible to deposit higher analogues of graphite subunits (e.g. compound **17**) on to a solid substrate using high vacuum vapour deposition technique [25b].

We have also shown that the self-assembly of 5-alkoxyisophthalic acid can be manipulated to give various morphologies at the interface. For example, 5-octadecyloxyisophthalic acid (Fig. 1) forms a layer-type morphology on a graphite substrate (Fig. 4, left) and this structure can be modified by adding an excess of a bifunctional hydrogen bond acceptor such as pyrazine to the supernatant solution (Fig. 4, right).[34]

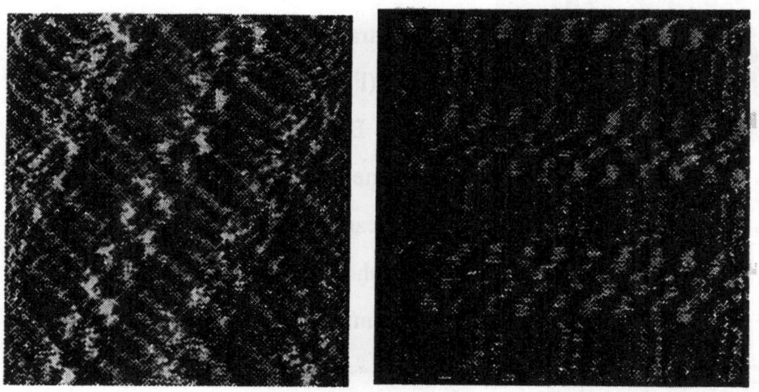

Figure 4: STM images of C16ISA (left) and C16ISA + Pyrazine (right)

Both these structures observed on graphite (2D) exhibit a remarkable resemblance to the structures observed in the single crystal lattice (3D).

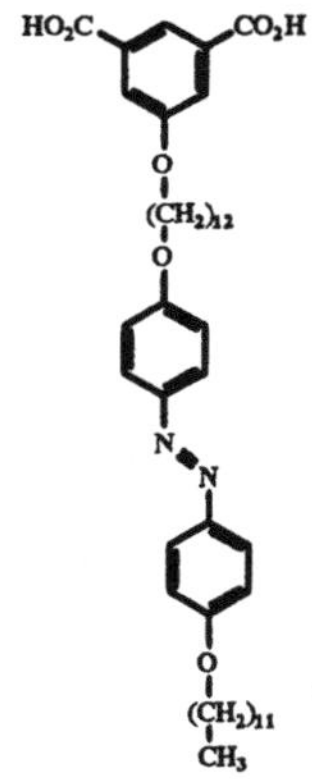

It should also be possible to address these supramolecular architectures derived from simple molecular building blocks by means of external stimuli such as light or electrons. This is important in establishing a way of storing information (writing) and observing it (reading). Our preliminary results indicate that it is possible to use light to induce changes such as or polymerisation of the monolayers on graphite.

Figure 5: STM images of trans isomers of ***C12AZOC12ISA.***

The first example, **C12AZOC12IS** deals with the formation of monolayers from an isophthalic acid derivative **(C12AZOC12ISA)** containing an azobenzene moiety [35].

Here the trans- isomer can be easily imaged by STM (Fig. 5) and the monolayer is very stable. By using light, a reversible photoisomerisation of the molecules in solution can be achieved. After the light induced photoisomerisation of trans- to cis isomers, multiple domains were observed on the surface. This implies that domains from the trans-isomer co-exist with those of the cis-isomer (Fig. 6). The reasons for this domain formation and the molecular reorganisation in the monolayer during

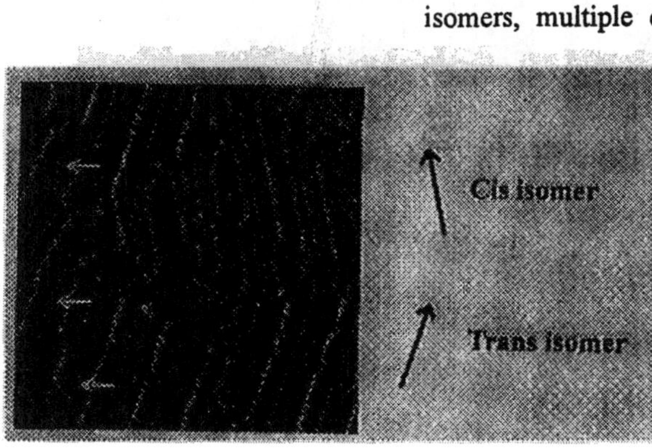

Figure 6: STM image of the monolayer of C12AZOC12ISA after phtotoisomerisation.

photoisomerisation are not clear and are under investigation.

3. Conclusions

As discussed above, it is possible to synthesise molecular building blocks whose properties can be tuned according to need. We have also employed a combination of the synthetic approach and the self-assembly concept to achieve structural control of the 2D-patterns. This, in turn, allows us to induce processibility and characterisability for the desired materials.

4. Acknowledgements

We thank the Bundesministerium für Forschung und Technologie and the Siemens AG. for financial support. S.V. and V. S. I. thank the Max-Planck-Institute for Polymer Research for financial support. The authors acknowledge stimulating interactions with the reasearch groups of Professors F. C. De Schryver and J. P. Rabe and help from all the collaborators whose names appear in the original publications.

References

1) a) G. M. Whitesides, J. P Mathias, C. T. Seto, *Science*, **1991**, *254*, 1312.; b) J. M. Lehn, *Angew. Chem. Int. Ed. Engl.*, **1988**, *27*, 89. c) G. A. Ozin, *Adv. Mater.*, **1992**, *4*, 612.

2) a) C. R. Martin, *Science*, **1994**, *266*, 1961. b) A. K. Chetham, *Science*, **1994**, *264*, 794. c) G. Wegner, *Mol. Cryst. Liq. Cryst.*, **1993**, *235*, 1. d) J. M. Lehn, *Macromol. Symp.*, **1993**, *69*, 1.

3) a) M. J. Sailer, C. L. Curtis, Adv. Mater., **1994**, *6*, 688.; b) A. M. Glass, *Science*, **1987**, *235*, 1003., c) F. L Carter, Molecular Electronics Devices I, Dekker New York, **1982** (I) and **1987** (II)., d) J. W. Thackeray, M. S. Wrighton, *J. Phys. Chem.*, **1986**, *90*, 6674.e) J. H. Burroughs, C. A. Jones, R. H. Friend, *Nature*, **1988**, *335*, 137. f) J. S. Miller, *Adv. Mater.*, **1990**, *2*, 378 (part A).495 (part B).

4) a) M. P. O'Neil, M. P. Niemczyk, W. A. Svec, D. Gosztola, G. L. Gaines III, M. R. Wasielewski, *Science*, **1992**, *257*, 63., b) J. J. Hopfield, J. N. Onuchic, D. N. Beratan, *Science*, **1988**, *241*, 817., c) D. A. Parthenopoulos, P. M. Rentzepis, *Science*, **1989**, *245*, 843.

5) S. Hunter, F. Kiamilev, S. Esener, D. A. Parthenopoulos, P. M. Rentzepis, *Appl. Opt.*, **1990**, *29*, 2058.

6) a) A. Aviram, M. A. Ratner, *Chem. Phys. Lett.*, **1974**, *29*, 277. b) M. D. Ward, *Chemistry and Industry*, **1996**, 569 and the references cited therein.

7) a) J. M. Tour, *Chem. Rev.*, **1996**, *96*, 537.b) M. Baumgarten, U. Bunz, U. Scherf, K. Müllen, *NATO ASI Series*, **1994**. c) T. Bein, (Edr.), Supramolecular Architecture, ACS Symposium Series 499, American Chemical Society, Washington, DC, **1992**, d) J. M. Tour, *Trends in Polym. Sci.*, **1994**, *2*, 332, e) J. S. Moore, J. Zhang, Z. Wu, D. Venkataraman, S. Lee, *Macromol. Symp.*, 77, 1994, 295. f) M. Baumgarten, K. Müllen, *Topics in Curr. Chem.*, **1994**, *169*, 1.

8) a) J. M. Tour, J. J. S. Lamba, *Macromol. Symp.*, **1994**, *77*, 389 and the references cited therein. b) D. L. Gin, V. P. Conticello, *TRIP*, **1996**, *4*, 217., c) J. Kido, *ibid*, **1994**, *2*, 350., d) G. Grem, G. Leditzky, B. Ullrich, G.

Leising, *Adv. Mater.*, **1992**, *4*, 36., e) G. Grem, V. Martin, F. Mehdadi, C. Paar, J. Stampfl, J. Sturm, S. Tasch, G. Leising, *Synth. Met.*, **1995**, *71*, 2193.

9) N. Miyaura, T. Yanags, A. Suzuki, *Synth. Commun.*, **1981**, *11*, 513.

10) a) W. Heitz, *Chem. Ztg*, **1986**, *110*, 385. b) M. Rehahn, A. -D. Schlüter, G. Wegner, W. T. Feast, *Polymer*, **1989**, *30*, 1060., c) J. K. Kallitsis, M. Rehahn, G. Wegner, *Makromol. Chem.*, **1992**, *193*, 1021., d) M. Rehahn, A. -D. Schlüter, G. Wegner, *Makromol. Chem.*, **1990**, *191*, 1991.

11) T. F. McCarthy, H. Witteler, T. Pakula, G. Wegner, *Macromolecules*, **1995**, *28*, 8350.

12) M. Kreyenschmidt, M. Baumgarten, N. Tyutyulkov, K. Müllen, *Angew. Chem.*, **1994**, *106*, 2062; *Angew. Chem. Int. Ed. Engl.*, *33*, 1957.

13) U. Scherf, K. Müllen, (1991); *Makromol. Chem. Rapid. Commun.*, *12*, 489.

14) P. Hickl, M. Ballauff, U. Scherf, K. Müllen, *Macromolecules*, **1996**, Submitted.

15) J. Stampfl, W. Graupner, G. Leising, U. Scherf, *J. Lumin.*, **1995**, *63*, 117.

16) a) J. Grüner, H. F. Wittmann, P. J. Hamer, R. H. Friend, J. Huber, U. Scherf, K. Müllen, S. C. Morati, A. B. Holmes, *Synth. Met.*, **1994**, *67*, 181. b) G. Grem, C. Paar, J. Stampfl, G. Leising, J. Huber, U. Scherf, *Chem. Mater.*, **1995**, *7*, 2. c) J. Huber, K. Müllen, J. Salbeck, H. Schenk, U, Scherf, T. Stehlin, R. Stern, *Acta Polymer.*, **1994**, *45*, 244.

17) J. Grimme, M. Kreyenschmidt, F. Uckert, K. Müllen, U. Scherf, *Adv. Mater.*, **1995**, *7*, 292.

18) a) F. Coat, C. Lapinte, *Organometallics*, **1996**, *15*, 477., b) G. Routh, H. Fischer, *ibid*, **1996**, *15*, 1139., c) L. M. Tolbert, et al, *J. Am. Chem. Soc.*, **1995**, *117*, 12891., d) T. Bartik, et al, *Angew. Chem. Int. Ed. Engl.*, **1996**, *35*, 414.

19) J. M. Tour, L. R. Jones II, D. L. Pearson, J. J. S. Lamba, T. P. Burgin, G. M. Whitesides, D. L. Allara, A. N. Parikha, S. V. Atre, *J. Am. Chem. Soc.*, **1995**, *117*, 9529.

20) V. Francke, K. Müllen, Unpublished results.

21) a) K. Sonoshigara, Y. Tohda, N. Hagihara, *Tetrahedron Lett.*, **1975**, 4467. b) S. Takahashi, K. Kuroayama, K. Sonoshigara, N. Hagihara, *Synthesis*, **1980**, 628., c) U. Scherf, K. Müllen, *Synthesis*, **1992**, 23., d) R. H. Grubbs, D. Kratz, *Chem. Ber.*, **1993**, *126*, 149.

22) D. Adam, P. Schuhmacher, J. Simmerer, L. Häussling, K. Siemensmeyer, K. H. Etzbach, H. Ringsdorf, D. Haarer, *Nature*, **1994**, *371*, 141.

23) A. Stabel, P. Herwig, K. Müllen, J. P. Rabe, *Angew. Chem.*, **1995**, *107*, 1768, *Angew. Chem. Int. Ed. Engl.*, **1995**, *34*, 1609.

24) P. Herwig, C. Kayser, H. W. Spiess, K. Müllen, *Adv. Mater.*, **1996**, *8*, 510.

25) a) M. Müller, H. Mauermann-Düll, M. Wagner, V. Enkelmann, K. Müllen, *Angew. Chem.*, **1995**, *107*, 1751; *Angew. Chem. Int. Ed. Engl.*, *34*, 1583. b) M. Müller, J. Petersen, R. Strohmeier, C. Günther, N. Karl, K. Müllen, *Angew. Chem. Int. Ed. Engl.*, **1996**, *35*, 886.

26) V. S. Iyer, K. Müllen, *et al*, Unpublished results.

27) a) G. M. Whitesides, J. P. Mathias, C. T. Seto, *Science*, **1991**, *254*, 1312, b) J. P. Mathias, E. E. Simanek, C. T. Seto, G. M. Whitesides, *Macromol. Symp.*, **1994**, *77*, 157. c) C. T. Seto, J. -P. Mathias, G. M. Whitesides, *J. Am. Chem. Soc.*, **1993**, *115*, 1321. d) M. C. Etter, *Acc. Chem. Res.*, **1990**, *23*, 120. e) G. F. Persico, J. D. Wuest, *J. Org. Chem.*, **1993**, *58*, 95. f) S. C. Zimmerman, B. F. Duerr, *J. Org. Chem.*, **1992**, *57*, 2215.

28) D. Philp, J. F. Stoddart, *Angew. Chem. Int. Ed. Engl.*, **1996**, *35*, 1154, *Angew. Chem.*, **1996**, *108*, 1242 and references cited therein.

29) a) S. Valiyaveettil, V. Enkelmann, K. Müllen, *J. Chem. Soc. Chem.Commun.* **1994**, 2097. b) V. Enkelmann, S. Valiyaveettil, G. Moessner, K. Müllen, *Supramol. Sci.*, **1995**, *2*, 3. c) S. Valiyaveettil, V. Enkelmann, G. Moessner, K. Müllen, *Mcromol. Symp.*, **1996**, *102*, 165.d) M. Pfaadt, G. Moessner, D. Pressner, S. Valiyaveettil, C. Boeffel, K. Müllen, H. W. Spiess, *J. Mater. Chem.*, **1995**, *5*, 2265. e) M. H. P. van Genderen, M. Pfaadt, V. Macho, S. Valiyaveettil, H. W. Spiess, *Ber. Bunsenges. Phys. Chem.*, **1996**, 100, 562.

30) S. Valiyaveettil, K. Müllen, Unpublished results.

31) a) A. J. Dann, H. Hoshi, Y. Maruyama, *J. Appl. Phys.*, **1990**, *67*, 1371, 1845. b) H. Yanagi, M. Ashida, J. Elbe, D. Wörnhle, J. Phys. Chem., **1990**, *94*, 7056., c) A. Schmidt, T. J. Schuerlein, G. E. Collins, N. R. Armstrong, *J. Phys. Chem.*, **1995**, *30*, 11770 and the references cited therein.

32) a) L. L. Soethout, H. Van Kempen, G. F. A. Van De Walle, *Adv. Electron. Phys.*, **1990**, *79*, 153., b) X. L. Wu, C. M. Lieber, *Prog. Inorg. Chem.*, **1991**, *39*, 431.

33) M. Pomerantz, A. Aviram, R. A. McCorkle, L. Li, A.G. Schrott, *Science*, **1992**, *255*, 1115 and references cited therein.

34) K. Eichhorst-Gerner, A. Stabel, D. Declerq, J. P. Rabe, G. Moessner, S. Valiyaveettil, V. Enkelmann, K. Müllen, *Angew. Chem.*, **1996**, *108*, 1599., *Angew. Chem. Int. Ed. Engl.*, **1996**, *35*, 1492.

35) P. Vanoppen, P. C. M. Grim, M. Rucker, S. De Feyter, G. Moessner, S. Valiyaveettil, K. Müllen, F. C. De Schryver, Submitted to *J. Am. Chem. Soc.*, **1996**.

SYNTHESIS OF MOLECULAR SCALE WIRES AND ALLIGATOR CLIPS

DARREN L. PEARSON, LEROY JONES II, JEFFRY S. SCHUMM AND JAMES M. TOUR*
Department of Chemistry and Biochemistry
University of South Carolina
Columbia, SC 29208, USA

Future computational system will likely consist of logic devices that are ultra dense, ultra fast, and molecular-sized.[1-3] The slow step in existing computational architectures is not usually the switching time, but the time it takes for an electron to travel between devices. By using molecular scale electronic interconnects,[4] the transmit times could be minimized, resulting in computational systems that operate at far greater speeds than is presently attainable from conventional patterned architectural arrays.[1] There is another technical advantage that might also be gained from molecular scale devices. A powerful computational system present utilizes approximately 10^{10} silicon-based devices. If devices were to be based upon single molecules, using routine chemical syntheses, one could prepare over 10^{23} devices in a single reaction flask! Though the task of addressing large arrays of ordered molecular scale devices is presently unattainable, the potential is certainly enough to maintain current and future interests.

Though it is well-documented that bulk conjugated organic materials can be semiconducting or even conducting when doped,[5] we only recently determined how thiol-ended rigid rod conjugated molecules orient themselves on gold surfaces,[6] and how to could record electronic conduction through single undoped conjugated molecules that are end-bound onto a metal probe surface.[7] Here we describe the synthetic details for the formation of soluble conjugated oligomers, potential molecular scale wires, by a rapid iterative divergent/convergent doubling approach.[8,9] Additionally, the syntheses and attachments of protected thiol moieties to one or both ends of the oligomers are presented. These thiols serve as molecular scale alligator clips for adhesion of the molecular scale wires to the gold probes.[6,7]

There has been considerable recent effort to prepare large conjugated molecules of precise length and constitution.[10] Our approach to these compounds maintains several key features that make it well-suited for the requisite large molecular architectures needed for molecular scale electronics studies. Specifically, the route involves (1) a rapid construction method that permits doubling molecular length at each coupling stage to

C. Joachim and S. Roth (eds.), Atomic and Molecular Wires, 81-87.

afford an unbranched 100 Å oligomer, the approximate size of present nanopatterned probe gaps,[1] (2) an iterative approach so that the same high yielding reactions can be used throughout the sequence, (3) the syntheses of conjugated compounds that are semiconducting in the bulk, (4) products that are stable to light and air so that subsequent engineering manipulations will not be impeded, (5) products that could easily permit independent functionalization of the ends to serve as molecular alligator clips that are required for surface contacts to metal probes, (6) products that are rigid in their frameworks so as to minimize conformational flexibility yet containing substituents for maintaining solubility and processability, (7) alkynyl units (cylindrically symmetric) separating the aryl units so that ground state contiguous π-overlap will be minimally affected by rotational variations, (8) molecular systems that do not have degenerate ground state resonance forms and are thus not subject to Peierls distortions,[5] and finally, (9) products that serve as useful models for the understanding of bulk polymeric materials.[5]

The iterative divergent/convergent approach is outlined below. A batch of monomer

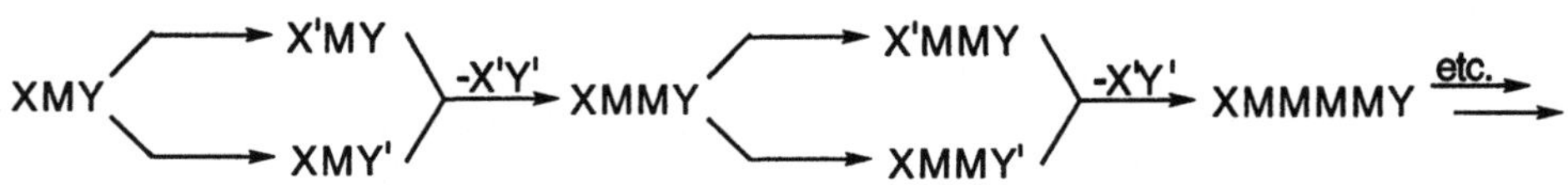

material M, with inactive end groups X and Y, is divided into two portions. In one portion, the end group X is activated by conversion to X'. In the second portion, Y is activated by conversion to Y'. The two portions are then brought back together to form the dimer XMMY with loss of X'Y'. Since the same end groups that were present in the monomer are now present in the dimer, the procedure can be repeated with a doubling of molecular length at each iteration. The advantages of this approach are that the molecular length grows rapidly, at a rate of 2^n were n = the number of iterations, and incomplete reactions yield unreacted material that is half the size of the desired compound. Thus, purification at each step is far simpler since separation involves, for example, an octamer from a 16-mer. This iterative divergent/convergent approach is therefore particularly attractive.

The specific iterative divergent/convergent synthetic approaches are outlined in Schemes 1 and 2. Notice how the oligomer lengths double after each iteration to rapidly form the desired 16-mers. These compounds have been characterized by UV-Vis, ^{1}H NMR, ^{13}C NMR, mass spectrometry, and in some cases, combustion analyses. While the tetrameric and octameric thiophene systems afforded molecular ions by direct exposure via electron impact mass spectrometry (MS), neither this method nor FAB or electrospray MS sufficed for obtaining a molecular ion of the 16-mer thiophene system. However, matrix assisted laser desorption MS (MALDI-MS) did afford an M+1 peak for the 16-mer (sinapinic acid matrix, positive ion mode). Similarly, for the phenylene system, the lower molecular weight oligomers could be characterized by direct exposure via electron impact MS. In order to obtain MS data on the soluble 3-ethylheptyl containing octamer and 16-mer, it was again necessary to use matrix assisted laser

Scheme 1

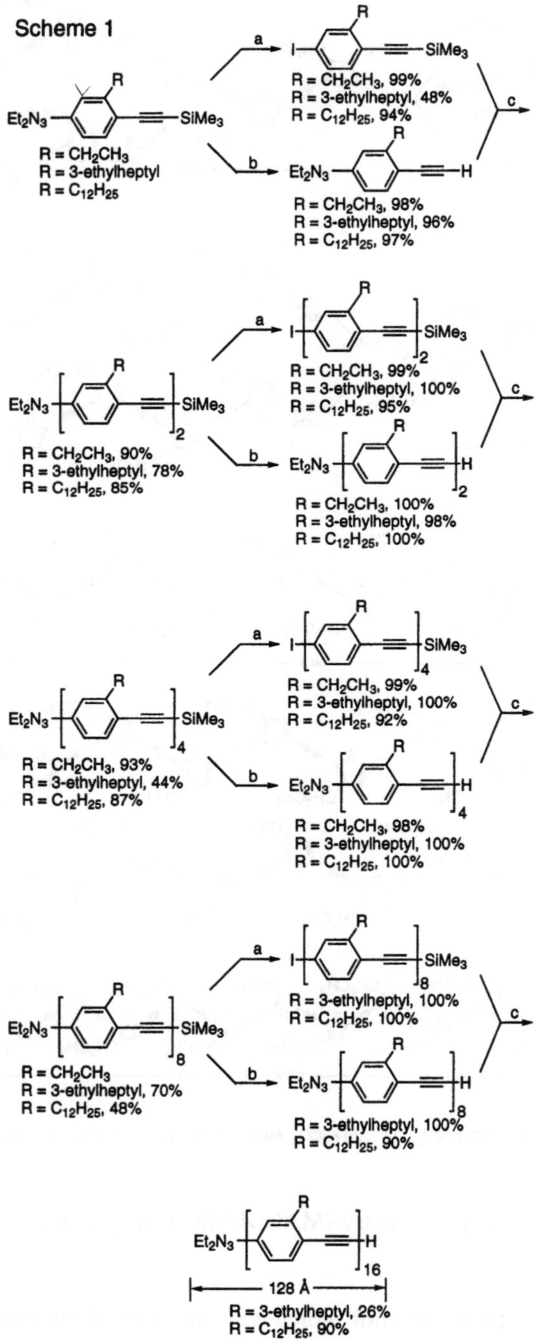

Reagents: a. MeI as solvent, 120°C in a screw cap tube. b. K_2CO_3, MeOH, 23°C or n-Bu_4NF, THF, 23°C c. $Pd(dba)_2$ (5 mol %), CuI (10 mol %), PPh_3 (20 mol %), i-Pr_2NH/THF (1:5), 23°C.

Scheme 2

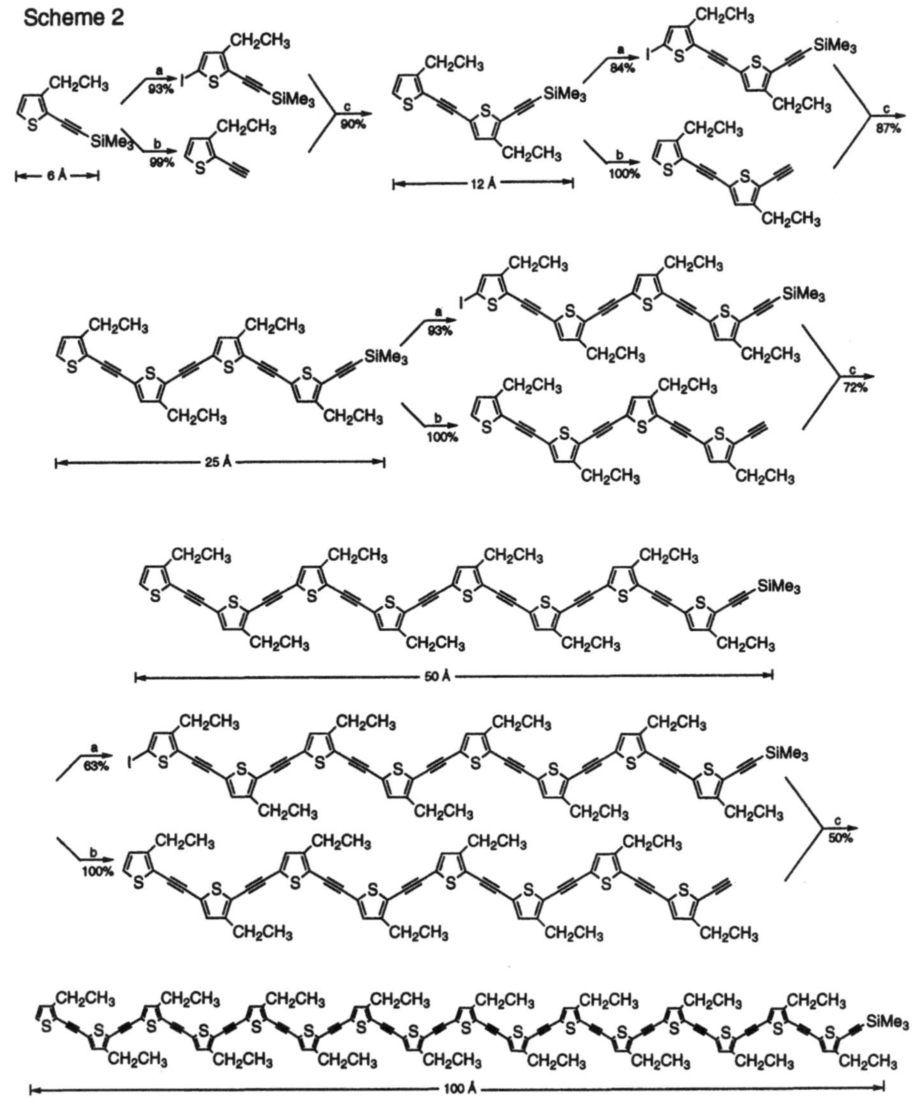

Reagents: (a) LDA, Et_2O, -78° to 0°C then I_2, -78°. (b) K_2CO_3, MeOH, 23°C. (c) $Cl_2Pd(PPh_3)_2$ (2 mol %), CuI (1.5 mol %), THF, i-Pr_2NH, 23 °C.

desorption/ionization mass spectrometry (MALDI-MS) using a sinapinic acid matrix, positive ion mode.

We then sought to affix protected thiol moieties to the ends of the oligomers. These thiols were used as molecular alligator clips for adhesion to gold probe surfaces.[6,7] In some cases, we prepared monothiol-terminated systems for adhesion of these oligomers to single gold surfaces. In other cases, we prepared the α,ω-difunctionalized systems for adhesion between two gold probe surfaces (see below).

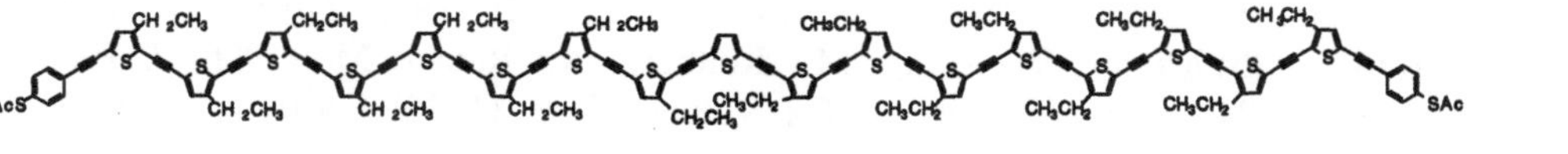

These are the macromolecules that we are presently using in our attempts to bridge proximal gold probes. Since the potential molecular wires must bridge two gold electrodes, the question as to how these rigid rod difuctional oligomers will order on gold surfaces comes into question. For example, will the oligomers bridge the gold-gold gap, or will they reside nearly parallel to the surface of the gold by either dithiol or aromatic adsorption to the gold surfaces? By making self-assembled monolayers on gold surfaces, we have demonstrated using ellipsometry, XPS, and grazing angle IR measurements, that the rigid rod systems predominantly stand perpendicular to the surface; the thiol groups dominating the adsorption sites on the gold. Even when the oligomers were α,ω-dithiol-substituted, the rigid molecules tended to stand on end as judged by the ellipsometric thickness of the adsorbate layer.[6a]

We found that the terminal aromatic thiols were most difficult to manipulate since they are unstable, undergoing disulfide formation in the presence of oxygen, often resulting in insoluble oligomers. However, the acetyl-protected thiols were resilient enough for manipulations in air, yet they could be readily hydrolyzed with NH_4OH, in situ, once exposed to the gold probe surfaces.[6a] This provided a convenient method for carrying out the depositions under conditions that were not rigorously oxygen-free; a condition common in most electronics fabrication facilities.

In summary, we outlined the rapid syntheses of conjugated oligomers via an iterative divergent/convergent approach. At each stage in the iteration, the length of the framework doubled. Only three sets of reaction conditions were needed for the entire iterative synthetic sequence: an iodination, a protodesilylation, and a Pd/Cu-catalyzed cross coupling. Convergent attachment of the complementary thiol end groups, protected as thioester moieties, was achieved. These serve as binding units for adhesion to gold surfaces. The rigid rod conjugated oligomers may act as molecular wires in molecular scale electronic devices.

Acknowledgments. We thank the Defense Advanced Research Projects Agency for support of this work.

References and Notes

1. For some recent background work on molecular scale electronics, see: (a) Aviram, A., Ed. (1992) *Molecular Electronics: Science and Technology,* Confer. Proc. No. 262, American Institute of Physics: New York. (b) Miller, J. S. (1990) *Adv. Mater.* 2, 378, 495, 601. (c) Kirk, W. P.; Reed, M. A., Eds. (1992) *Nanostructures and Mesoscopic Systems*, Academic: New York. (d) Birge, R. R.,

Ed. (1991) *Molecular and Biomolecular Electronics*, Advances in Chemistry Series 240, American Chemical Society.

2. For some theoretical considerations on molecular scale wires, see: (a) Samanta, M. P.; Tian, W.; Datta, S.; Henderson, J. I.; Kubiak, C. P. (1993) *Phys. Rev. B*, *53*, R7626. (b) Mujica, V.; Kemp, M.; Roitberg, A.; Ratner. M. (1996) *J. Phys. Chem.*, *104*, 7296. (c) Joachim, C.; Vinuesa, J. F. (1996) *Europhys. Lett.*, *33*, 635.
3. For some recent background work on the formation of molecular-based transporters and devices, see: (a) Grosshenny, V.; Harriman, A.; Ziessel R. (1995) *Angew. Chem. Int. Ed. Engl.*, *34*, 1100. (b) Langler, L.; Stockman, L; Heremans, J. P.; Bayor, V.; Olk, C. H.; Van Haesendonck, C.; Bruynseraede, Y; Issi, J. -P. (1995) *Synth. Met.*, *70*, 1393. (c) Pascual, J. I.; Méndez, J; Gómez-Herrero, J; Baró, A. M.; Garcia, N; Landman, U.; Luedtker, W. D.; Bogachek, E. N.; Cheng, H. -P. (1995) *Science 267*, 1793. (d) Purcell, S. T.; Garcia, N.; Binh, V. T.; Jones II, L; Tour, J. M. (1994) *J. Am. Chem. Soc. 116*, 11985. (e) Martin, C. R. (1994) *Science 266*, 1961. (f) Seth, J.; Palaniappan, V.; Johnson, T. E.; Prathapan, S.; Lindsey, J. S.; Bocian, D. F. (1994) *J. Am. Chem. Soc. 116*, 10578. (g) Gust, D. (1994) *Nature 372*, 133. (h) Wu, C.; Bein, T. (1994) *Science 266*, 1013. (i) Wagner, R. W.; Lindsay, J. S.; Seth, J.; Palaniappan, V.; Bocain, D. F. (1996) *J. Am. Chem. Soc. 118*, 3996. (j) Sailor, M. J.; Curtis, C. L. (1994) *Adv. Mater. 6*, 688. (k) Brigelletti, F.; Flamigni, L; Balzani, V.; Collin, J.; Sauvage, J.; Sour, A.; Constable, E. C.; Thompson, A. M. W. C. (1994) *J. Am. Chem. Soc. 116*, 7692. (l) Wu, C.; Bein, T. (1994) *Science 264*, 1757. (m) Moerner, W. E. (1994) *Science 265*, 46. (n) Sessler J. L.; Capuano, V. L.; Harriman, A. (1993) *J. Am. Chem. Soc. 115*, 4618. (o) Farazdel, A.; Dupuis, M.; Clementi, E.; Aviram. A. (1990) *J. Am. Chem. Soc. 112*, 4206. (p) Tour, J. M.; Wu, R.; Schumm. J. S. (1991) *J. Am. Chem. Soc. 113*, 7064. (q) Dai, H.; Wong, E. W.; Lieber, C. M. (1996) *Science 272*, 523. (r) Wu, R.; Schumm, J. S.; Pearson, D. L.; Tour, J. M. (In press) *J. Org. Chem.*
4. One of the ill-defined terms in the area of molecular electronics is simply the definition of "molecular electronics", since some authors refer to it as any molecular-based system such as a film or a liquid crystalline array. Other authors, including us, have preferred to reserve the term "molecular electronics" for single molecule tasks, such as single molecule-based transistors or single molecular wires. Due to this confusion, we have chosen here to follow the Petty *et al.* (*Introduction to Molecular Electronics,* Petty, M. C.; Bryce, M. R.; Bloor, D. Eds.; Oxford Univ. Press: New York, 1995) terminology by using two subcategories, namely "molecular materials for electronics" for bulk applications and "molecular scale electronics" for single molecule applications.
5. Skotheim, T. A., Ed. (1986) *Handbook of Conducting Polymers*, Dekker: New York.
6. (a) Tour, J. M.; Jones, L., II; Pearson, D. L.; Lamba, J. S.; Burgin, T.; Whitesides, G. W.; Allara, D. L.; Parikh, A. N.; Atre, S. (1995) *J. Am. Chem. Soc. 117*, 9529. (b) Dhirani, A.; Zehner, R. W.; Hsung, R. P.; Guyot-Sionnest, P., Sita, L. R. (1996) *J. Am. Chem. Soc. 118*, 3319.

7. Bumm, L. A.; Arnold, J. J.; Cygan, M. T.; Dunbar, T. D.; Burgin, T. P.; Jones, L., II, Allara, D. L.; Tour, J. M.; Weiss, P. S. (1996) *Science* *271*, 1705.
8. Schumm, J. S.; Pearson, D. L.; Tour, J. M. (1994) *Angew. Chem. Int. Ed. Engl.* *33*, 1360.
9. Pearson, D. L.; Schumm, J. S.; Tour, J. M. (1994) *Macromolecules* *27*, 2348.
10. Tour, J. M. (1996) *Chem. Rev.* *96*, 537.

SYNTHESIS OF CONJUGATED LADDER OLIGOMERS

A. GOURDON

Groupe Electronique Moléculaire
CEMES-CNRS UP8011
29, Rue Jeanne Marvig
BP 4347
31055 Toulouse (France)

Introduction

Considerable interest is being shown in the study of 1) long distance electron (and energy) transfer in homo and heterodinuclear metallic complexes and 2) electron transport through molecular wires between two nanoelectrodes [1]. The measurement of these properties is limited for 1) by the low intensity of the intervalence band in the near-IR which provides informations on the electronic coupling through the molecular wire and for 2) by the detection of the low intensity current between the two nanoelectrodes. As a consequence the molecular wires and their electronic coupling with the metallic ends must be optimized as much as possible. Although single electron transfer in 1) and bulk electron transport in 2) cannot be directly compared, for both experiments the bridging molecules must meet very similar requirements:

a) well defined length, which excludes polymers:

1) The study of long distance electron transfer in function of the bridge length requires the synthesis of a series of long (up to 50 Å) oligomers with various controlled lengths

2) Incorporation of a molecular wire between two nanoelectrodes requires the preparation of oligomers [2] significantly longer than the interelectrode gap (100 Å) to allow a good 'electrical' connection.

b) rigidity: the molecule must not be folded, which excludes molecules with conformers in which the overall length is not fixed as polythiophenes.

c) good electron transfer or transport properties. One of the key-point is a small homo-lumo gap by extended π-conjugation.

d) good electronic coupling with the metallic ends. Recents papers have emphasized the crucial importance of this point. For a better electron transfer, the organic part can be bonded to the metal atom by a metal to carbon bond for example [3, 4]. For electron transport 2, the molecule must be chemisorbed, lying flat on the surface with strong coupling between the molecule π-orbitals and the metal orbitals [5]. It must be stressed that the values of the electronic transparence can vary by 4 to 5 orders of magnitude

C. Joachim and S. Roth (eds.), Atomic and Molecular Wires, 89-98.

depending upon the electrode-molecule coupling. It will therefore be absolutely necessary to get a very precise description of the geometry electrode-molecule-electrode to extract meaningful informations from the probably erratic measured values of the conductance.

e) solublity for chemical processing and purifications.

f) chemical and photochemical stability which precludes, for example, the use of some long photosensitive polyenes

In order to avoid a loss of conjugation by rotation around single or triple bonds we have focused our synthetic efforts on fused aromatic ladders, mainly oligoquinoxalines and oligobenzoanthracenes.

Oligoquinoxalines

Quinoxalines can be obtained by condensation of aromatic α-diamines and α-diketones (Hinsberg condensation):

Although this reaction looks simple, the reaction mechanism is not fully understood and involves (at least) eight atoms with elimination of two molecules of water. The synthesis of oligoquinoxalines proceeds by condensation of partially protected tetraminobenzene and tetraketopyrene, to avoid uncontrolled polymerizations. For example the *ca* 30 Å long **2** has been obtained according to the scheme 1. Condensation of the partially tosylated benzene tetramine with the tetraketopyrene, obtained by oxydation of pyrene in 5 steps, gives the compound **1**. Deprotection in acidic medium yields the tetramine derivative wich condenses with the 1,10-phenanthroline-4,5-dione (phendione) to give the poorly soluble ladder **2**. Attempts to improve the solubility by substitution of the benzene rings requires the synthesis of hexafunctionalized benzene rings which were found difficult to access. One alternative possibility is to use the anthracene instead of the benzene ring as starting building block. For instance, condensation of veratrole with substituted benzaldehyde provided 2,3,6,7 tetramethoxyanthracene with substituents at position 9 and 10 for solubility. Cleavage of the four methoxygroups by BBr_3 or pyriliumchloride gave the hydroquinone which, after oxidation to quinhydrone or dihydroxyanthraquinone, could be used as starting material for further condensation with aromatic diamines.

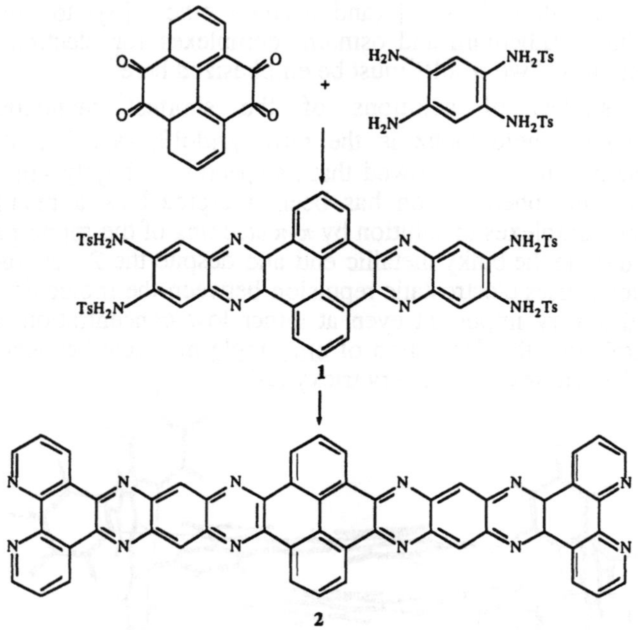

Scheme 1

An another possibility is to build up the bridging ligand from a soluble monomer precursor, and in particular to use the reaction of precoordinated ligands. This allows the preparation of soluble mononuclear and asymmetric binuclear complexes bridged by poorly soluble or even fully insoluble ligands [6] as outlined below:

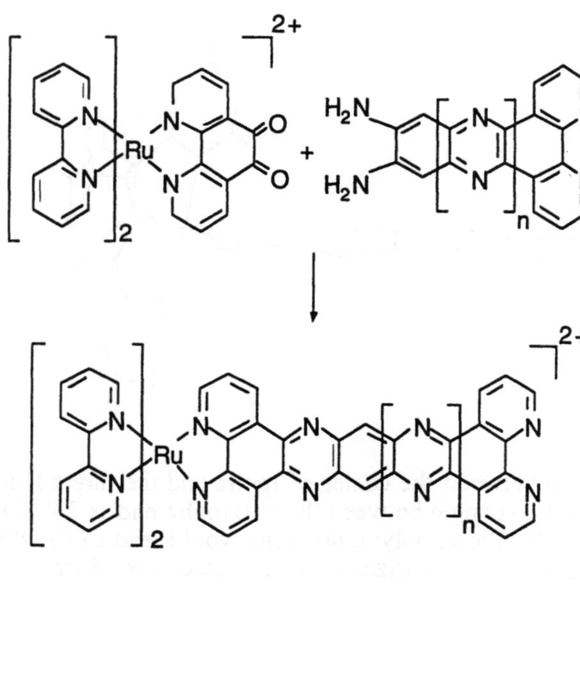

This approach has allowed us [7] (and recently others [8]) to prepare various very soluble ruthenium and osmium complexes for electron and energy transfer studies. Two results must be emphasized here:

- ^1H NMR studies of solutions of the shortest mononuclear $[Ru(bpy)_2(tpphz)]^{4+}$, where tpphz is the tetrapyrido[3,2-a:2',3'-c:3",2"-h:2"',3"'-j]phenazine (n = 0), showed that its spectra are highly sensitive to concentration. This phenomenon has been interpreted as a result of aggregation of the complexes in solution by π–π stacking of the tpphz parts. It indicates, that despite the bulky metallic end and despite the 2+ charge of the complex which causes electrostatic repulsion between the molecules, π–π stacking remains very important even at rather low concentration. It is therefore very likely that the deposition of <u>one single</u> molecule between the two nanoelectrodes will reveal as a very tricky task !

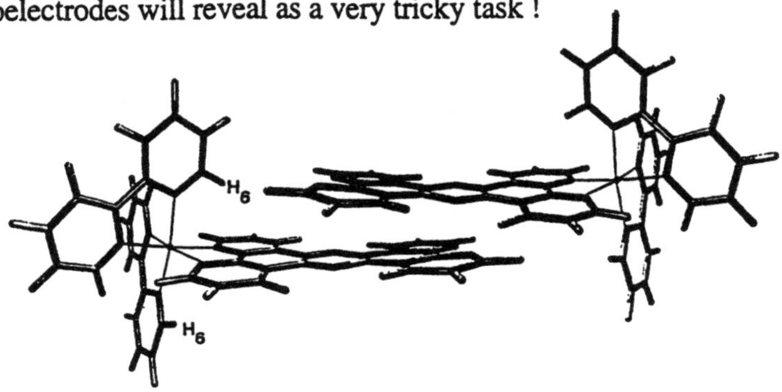

- the X-ray structure of the dinuclear complex $[Ru(bpy)_2(tpphz)Ru(bpy)_2]^{2+}$ has showned that, in the crystal, the ligand is at the same time bent and twisted as illustrated below:

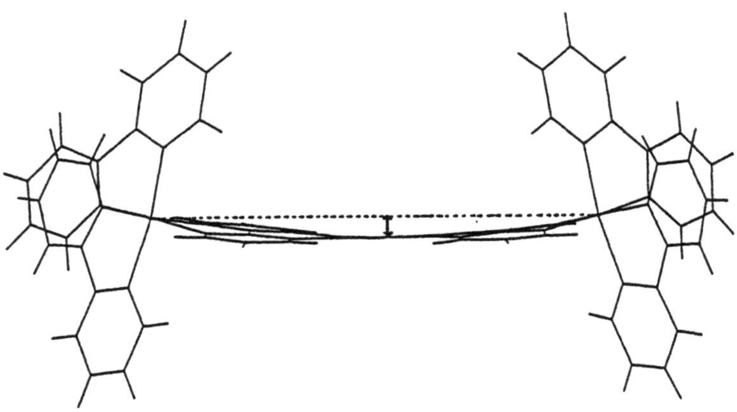

The maximum distance between the center of tpphz and the direct Ru-Ru vector is 0.57 Å and the twist angle between the two tpphz end is 18°. Such distortions applied to a 100 Å long polyquinoxaline would lead to the shape of a Möbius ring !. This point emphasizes an unexpected lack of rigidity of these polyaromatic systems, probably common to all the long conjugated oligomers described so far in the litterature [9]. It raises doubts on the real shape of long molecules chemisorbed or physisorbed on the bumpy surfaces of the nanoelectrodes and in the gap between these electrodes.

Oligobenzoanthracenes

Benzoanthracenes can be synthesized via Swager's electrophile-induced cyclisation of 2-(ethynyl)biphenyl [10]:

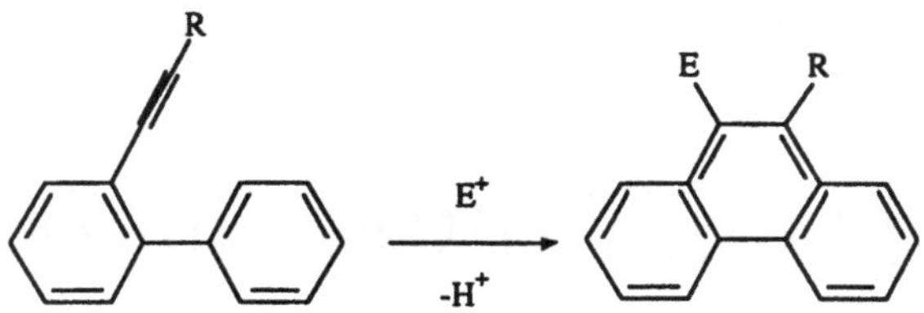

These biphenyls are obtained by Suzuki's coupling [11] of aromatic halides and boronic esters in presence of a palladium catalyst and a base. The synthesis of polybenzoanthracenes follows this scheme from poly(p-phenylene) substituted with acetylenic functions [from ref. 10]:

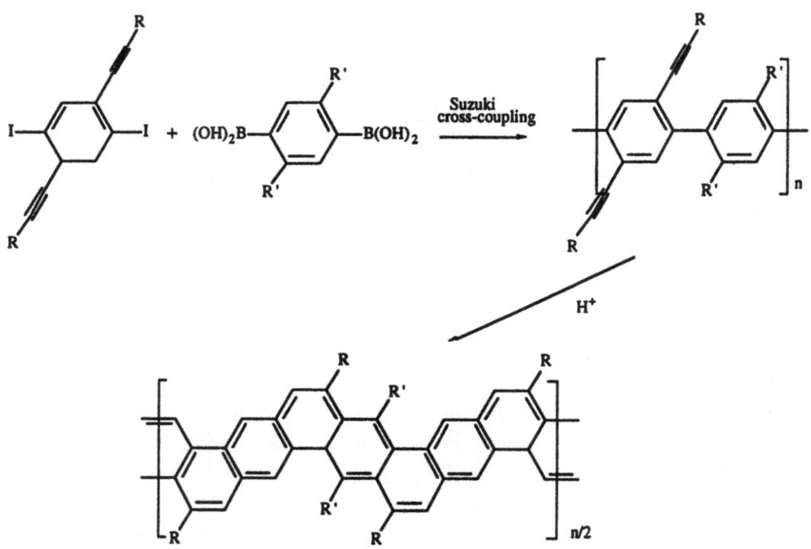

Controlled chain length requires a new synthetic route and different starting building blocks. We have developped a crossed-divergent convergent approach shown on scheme 2.

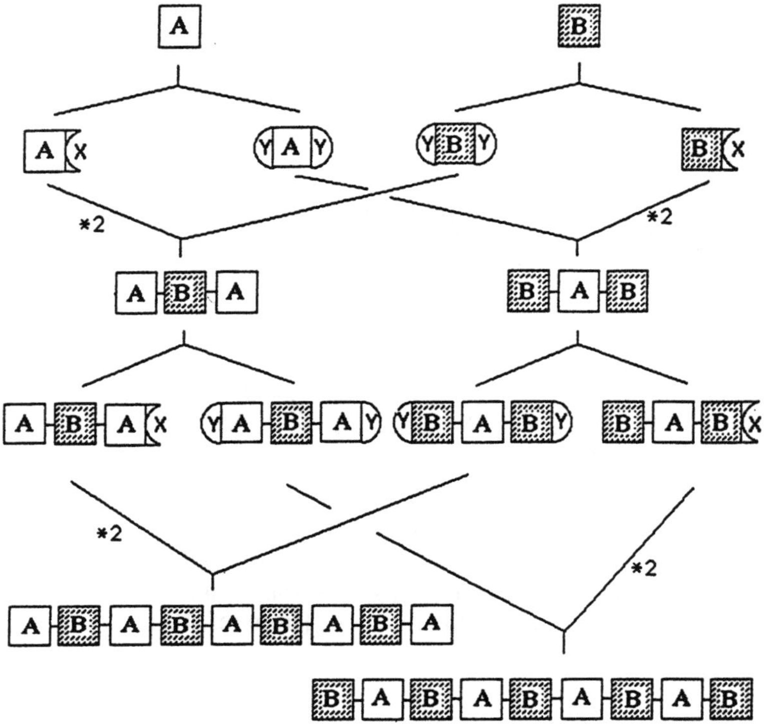

Scheme 2

In a first step, each of the two starting monomers A and B is mono- and di-functionalized to get the four species AX, YAY, BX and YBY. The cross-coupling of these monomers 2AX + YBY and 2BX + YAY gives respectively ABA and BAB with "elimination" of XY. The ABA and BAB terphenylenes are not very different from A and B and can be mono- and di-functionalized in a similar way to give the nonamers, and so on.

The main advantage of this crossed iterative divergent-convergent route is a very rapid increase of the molecule length wich is tripled at each coupling. The chain growth is 3^n instead of 2^n (n=number of iterations) for a linear iterative divergent-convergent method. For example, the synthesis of a 100 Å long poly(p-phenylene) requires only 3 coupling reactions instead of 5 in the case of a linear process. However this approach has a main drawback: the double Suzuki's couplings lead to symmetrical molecules A-L-A and B-L-B which must then be functionnalized at only one end. This monofunctionalization, which can be efficient for short molecules for which a first reaction at one end can prevent a second reaction at the other end, is rapidly less efficient as the chain growth. The reaction leads to statistical mixture of mono and difunctionalized species, which lowers the yields and causes purification problems.

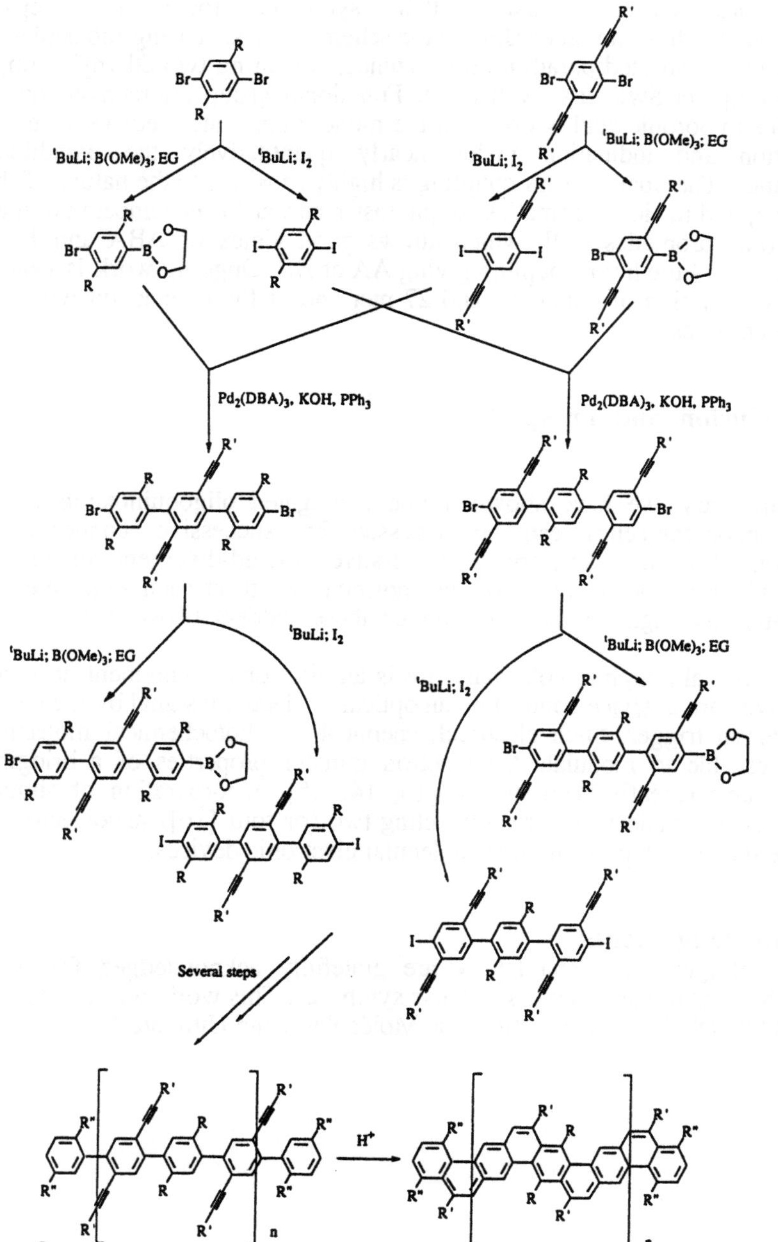

Scheme 3.

Reagents: Diiodination of the dibromo compounds: 2-fold excess of tBuLi, THF, -78 °C, I_2; yields 95-98%. Boronic esters: tBuLi (stoechio), THF, -78 °C; $B(OMe)_3$; HCl, chromat.; ethyleneglycol; yields: monomers: 72-80 %, terphenylene 35-40%. Cross-coupling: $Pd_2(DBA)_3$, KOH or Na_2CO_3, toluene or nitrobenzene, 70 °C, 12-15 h; yields: terphenylene: 60-70 %, nonaphenylene: 30 % (preliminary result). R = C_6H_{13}, R' = $PhOC_{12}H_{25}$, R" = CH_3.

We are currently using this synthetic route to prepare oligobenzoanthracenes according to the scheme 3. The starting molecules A and B are substituted paradibromobenzenes, B bearing two ethynyl groups for subsequent Swager's cyclisation. Functional group interconversion of bromine to boronic acid is done via the monolithium intermediate, whereas dilitiation and iodination yields nearly quantitatively the paradiiodo aromatics. The Suzuki cross-coupling is highly sensitive to the nature of the halogen, and the iodo aromatics couple faster and at lower temperature than the bromo aromatics [10], which allows good yields of ABA and BAB species with little homocoupling giving AA or BB. Ongoing work is geared at the preparation of nonamers and 27-mer and at the connection with the gold electrodes.

Conclusion and prospects

In summary, we have showned that conjugated oligoquinoxalines and oligobenzoanthracenes can be accessed by successive condensation reactions. We have developped an iterative crossed-divergent-convergent route which allows a tripling of the molecule length at each step. We are currently investigating the adhesion of these ladders between two gold probes.

However, although a molecular wire is an electronic component, it is not an active part, a <u>device</u> controlling an optical or electrical signal by means of an external trigger. Such electrical, chemical or photochemical molecular switches able to modulate the electron transfer properties of a bridging ligand have recently been studied [13, 14, 15]. Incorporation of similar switches in a molecular wire connecting two (or four [16]) nanoelectrodes will be the next step toward true molecular electronic devices.

Acknowledgements

Joseph Bolger and Elena Ishow are gratefully acknowledged for their contributions to various parts of these syntheses. This work was supported by the CNRS, the GDR Électronique Moléculaire and Ultimatech.

References

[1] Ward, M.D. (1996) Current developments in molecular wires, *Chem. Ind.*, 568-573.

[2] Tour, J.M. [1996] Conjugated molecules of precise length and constitution. Organic synthesis for the construction of nanoarchitectures, *Chem. Rev.* , **96**, 537-553.

[3] Patoux, C., Launay, J.P., Collin, J.P., Sauvage, J.P. and Chodorowski-Kimmes, S. Submitted to the J. Am. Chem. Soc.

[4] Collin, J.P., Lainé, P., Launay, J.P., Sauvage, J.P. and Sour, A. [1994] Pronounced electronic coupling in rigidly connected N,C,N-coordinated diruthenium complexes over a distance of up to 20 Å, *Angew. Chem., Int. Ed. Engl.*, **33**, 1775-1778.

[5] Joachim, C. and Vinuessa, J.F. (1996) Length dependence of the electronic transparence (conductance) of a molecular wire, *Europhys. Lett.* , **33**, 635-640.

[6] Bolger, J., Gourdon, A., Ishow, E. and Launay, J.-P (1995) Stepwise Syntheses of tpphz Mono and Dinuclear Ruthenium Complexes $[(bpy)_2Ru(tpphz)]^{2+}$ and $[(bpy)_2Ru(tpphz)Ru(bpy)_2]^{4+}$ (tpphz : tetrapyrido[3,2-a:2',3'-c:3", *J. Chem. Soc., Chem. Comm.* , 1799-1800.

[7] Bolger, J., Gourdon, A., Ishow, E. and Launay, J.-P. (1996) Mononuclear and Binuclear Tetrapyrido[3,2-a:2',3'-c:3",2"-h:2"',3"'-j]phenazine (tpphz) Ruthenium and Osmium Complexes, *Inorg. Chem.* **35,** 2937-2944.

[8] Wärnmark, K., Thomas, J.A., Heyke, O. and Lehn, J.M. (1996) Stereoisomerically controlled inorganic architectures: sythesis of enantio- and diastereo-merically pure ruthenium-palladium molecular rods from enathiopure building blocks, *Chem. Commun.*, 701-702.

[9] See for example :Llinger, N.L. and Pathiaseril, A. [1987] A force field for allenes and nonlinear acetylenes within the MM2 approximation, *J. Comput. Chem.*, **8**, 1225-1231.

[10] Goldfinger, M.B. and Swager, T.M. (1994) Fused Polycyclic aromatics via electrophile-induced cyclization reactions: application to the synthesis of graphite ribbons, *J. Am. Chem. Soc.* , **116**, 7895-7896.

[11] Miyaura, N. and Suzuki, A. (1995) Palladium-catalyzed cross-coupling reactions of organoboron compounds, *Chem. Rev.*, **95**, 2457-2483.

[12] See for example: Liess, P., Hensel, V. and Schlüter, A.-D. (1996) Oligiphenylene rods: a repetitive approach, *Liebigs Ann.*, 1037-1040.

[13] Hatzidimitriou, A.G., Gourdon, A., Devillers, J., Launay, J.-P., Mena, E. and Amouyal, E. (1996) Photoinduced Electron Transfer in Pentaamineruthenium(II) complexes of 1-(4-cyanophenyl)imidazole, *Inorg. Chem.* **35**, 2212-2219.

[14] Lainé, P., Marvaud,V., Gourdon, A., Launay, J.-P., Argazzi, R. and Bignozzi, C.-A. (1996) Electron transfer through norbornadiene and quadricyclane moieties as a model for molecular switching, *Inorg. Chem. 35,* 711-714.

[15] Gilat, S.L., Kawai, S.H. and Lehn, J.M. (1995) Light-triggered molecular devices: photochemical switching of optical and electrochemical properties in molecular wire type diarylethene species, *Chem. Eur.* , **1**, 275-284.

[16] Tour, J.M., Wu, R. and Schumm, J.S. (1991) Extended orthogonally fused conducting oligomers for molecular electronic devices, *J. Am. Chem. Soc.* , **113**, 7064-7066.

SYNTHESIS OF NANOWIRES ENCAPSULATED IN CARBON NANOTUBES BY THE ARC DISCHARGE METHOD*

H. PASCARD
Laboratoire des Solides Irradiés
CEA/CNRS URA 1380
Ecole Polytechnique,
91128, Palaiseau Cedex, France

* Work in collaboration with A. Loiseau (ONERA), N. Demoncy (CEA-ONERA), F. Willaime (CEA), P. Laplace (CEA), C. Laumonier (CEA), and in 1993-1994, C.Guerret, Y. Le Bouar, A. Tanguy, D. Weic (Ecole Polytechnique)

1. Introduction

After the discovery of a method to produce fullerenes in large quantities, based on the vaporization of graphite by electric arc discharge I1I, it was discovered the existence of carbon nanotubes as one of the products of the same experiment I2I. Since these discoveries, a great number of attempts have been made to fill carbon nanotubes with various elements or compounds, in order to elaborate nanowires encapsulated in carbon nanotubes. Several approaches have been used to introduce the elements : by capillarity, by chemical reaction and directly by electric arc discharge. In the capillary method, the filling of the tubes is obtained through capillary action in the presence of liquid metal near the opened tube ends I3I. In the chemical method, the encapsulation is made by opening the tubes and depositing the filling material using wet chemical techniques I4I. In the arc discharge method, the encapsulation is obtained by the formation of nanotubes in the plasma of an electric arc with the elements present in situ I5I.

C. Joachim and S. Roth (eds.), Atomic and Molecular Wires, 99-108.

Here, we present the method of synthesizing of nanowires encapsulated in carbon nanotubes by the electric arc discharge method.

2. Description of the reactor for nanowires synthesis

Figure 1 shows schematically the experimental set-up used for the synthesis of nanowires. The apparatus is a modified fullerene reactor, in which a composite anode is used. The reactor consists mainly of a vacuum chamber, a DC current generator, two water-cooled copper rods to supply the current, a graphite cathode, a composite graphite-metal anode, an electrode moving device and a water-cooled copper sleeve surrounding the plasma. The vacuum chamber is evacuated with a mechanical vacuum pump and then refilling with a gas (helium, nitrogen,..) to a static pressure between 0.1 and 0.8 bar. The DC generator is a DC arc welder supplying about 3 kW during the plasma experiment with a current of around 100 A and a voltage of around 30 V. The graphite electrodes are fixed on two water-cooled copper rods which supply the current. The cathode is a graphite rod 9 mm in diameter and 42 mm in length. The anode is a graphite rod 9 mm in diameter, 72 mm in length, with a hole 6 mm in diameter drilled to a depth of 38 mm and filled with a mixture of graphite and element powders.The element and the graphite powders are mixed by choosing a graphite/metal volume ratio of roughly the percolation threshold for the metal (70% graphite, 30% element). After packing the mixture, the volume of the anode hole is filled by approximately 15 % metal, 35% graphite and 50% empty space. One of the copper rods supplying the current is mounted on a mechanical moving device allowing manual adjustment of the distance between electrodes, typically from 1mm to 1 cm. The plasma region is surrounded by a water cooled copper cylinder about 15 cm in diameter.

3. Description of one nanowires-synthesis experiment

The synthesis is based on several physical and chemical effects : the electric-arc-discharge, the high-temperature-plasma, the consumption of the composite anode and the deposit on the cathode. A typical experiment has the following steps : - establishment of the current, - creation of the electric arc, - appearance of the plasma, - anode consumption

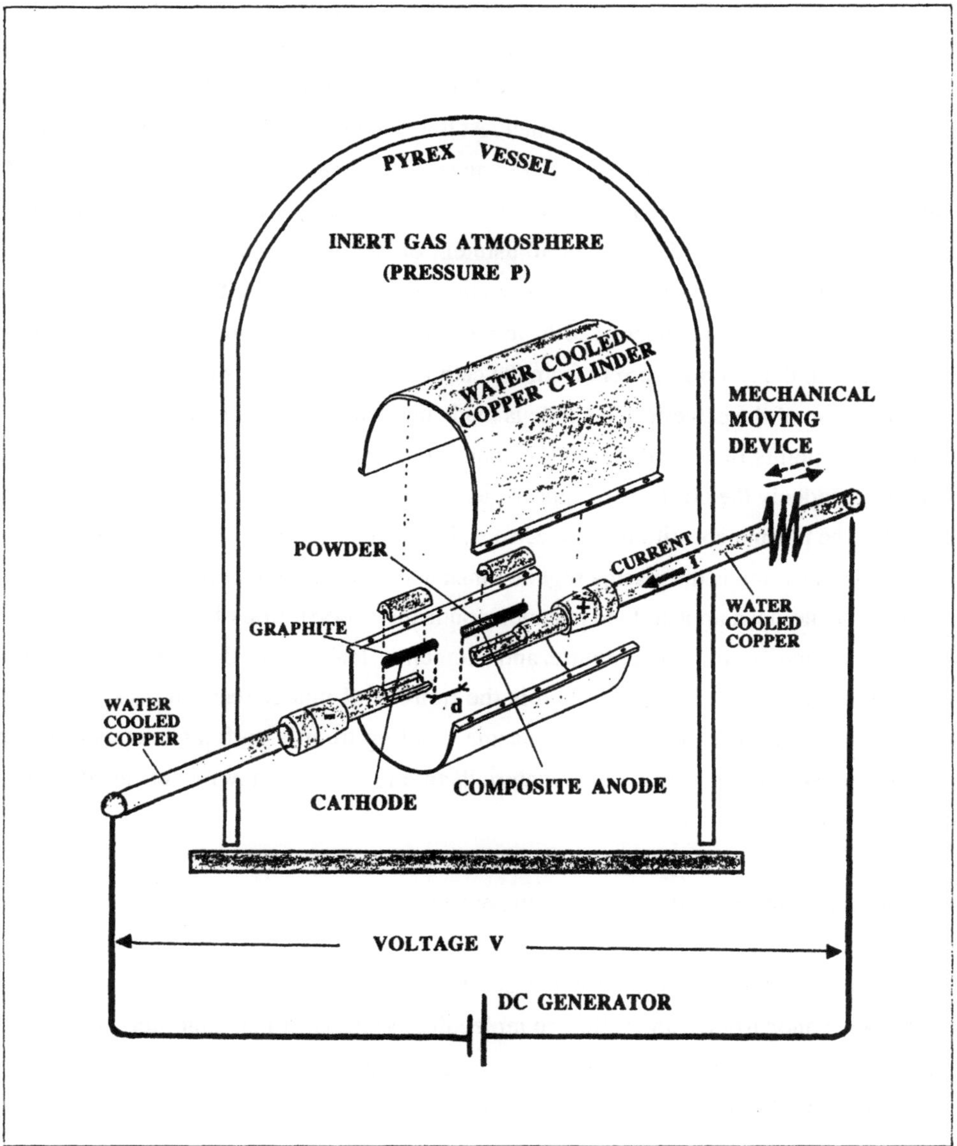

Figure 1. Schematic drawing of the electric-arc-discharge reactor for synthesis of nanowires encapsulated in carbon nanotubes.

by boiling of the metal and graphite, - transfer in the plasma of metal and carbon ions, - formation of various nanomaterials at different places in the reactor. Figure 2 shows schematically the different steps and figure 3 the evolutions of voltage and current versus time, with variable inter-electrode distance. At the beginning (t = 0), the electrodes are not in contact. The electric connection of the generator is made : the current I = 0 and the voltage V = 65 V. At the first step (t = 1 sec.), the mechanical contact between cathode and anode is established by manual adjustment of the inter-electrode distance. A high current of electrons is established. The voltage drops from 65 V to about 3 V and the current takes a high value about 120 A. Immediately, the anode temperature increases by heating due to the electron current from the cathode. At the second step (about t = 3 sec.), the separation of the cathode and anode is made manually by moving the anode. Locally, at different points of contact cathode-anode, the voltage is higher than the ionization voltage : suddenly the electric arc appears, and the voltage abruptly increases to about 20 to 30 V. The plasma constituted by electrons, ions,atoms (carbon , metal, helium), takes place. The third step is the plasma regime : the anode temperature becomes very high and two new phenomena appear. One phenomenon is the vaporization of the anode materials by melting and boiling of graphite and elements, leading to the composite-anode consumption. The other phenomenon is the transfer of matter (carbon ions, elements ions) from anode to cathode by the plasma, leading to an significant deposit on the cathode constituted of various nanomaterials among which nanowires encapsulated in carbon nanotubes.

4. Description of synthesized nanomaterials

Three kinds of nanomaterials are synthesized during the same experiment : carbon nanotubes, nanoparticles and fullerene molecules. Various carbon nanotubes are found mainly in the deposit on the cathode : empty graphitic nanotubes, filled graphitic nanotubes with different elements (pure elements, carbides,..). Various nanoparticles are found at different places in the reactor : nanocrystallites of elements encapsulated in graphitic planes, graphitic particles, amorphous carbon particles,.. Various fullerene molecules are found in the soot in the reactor : fullerenes, endohedral metallofullerenes, derivates of fullerenes...

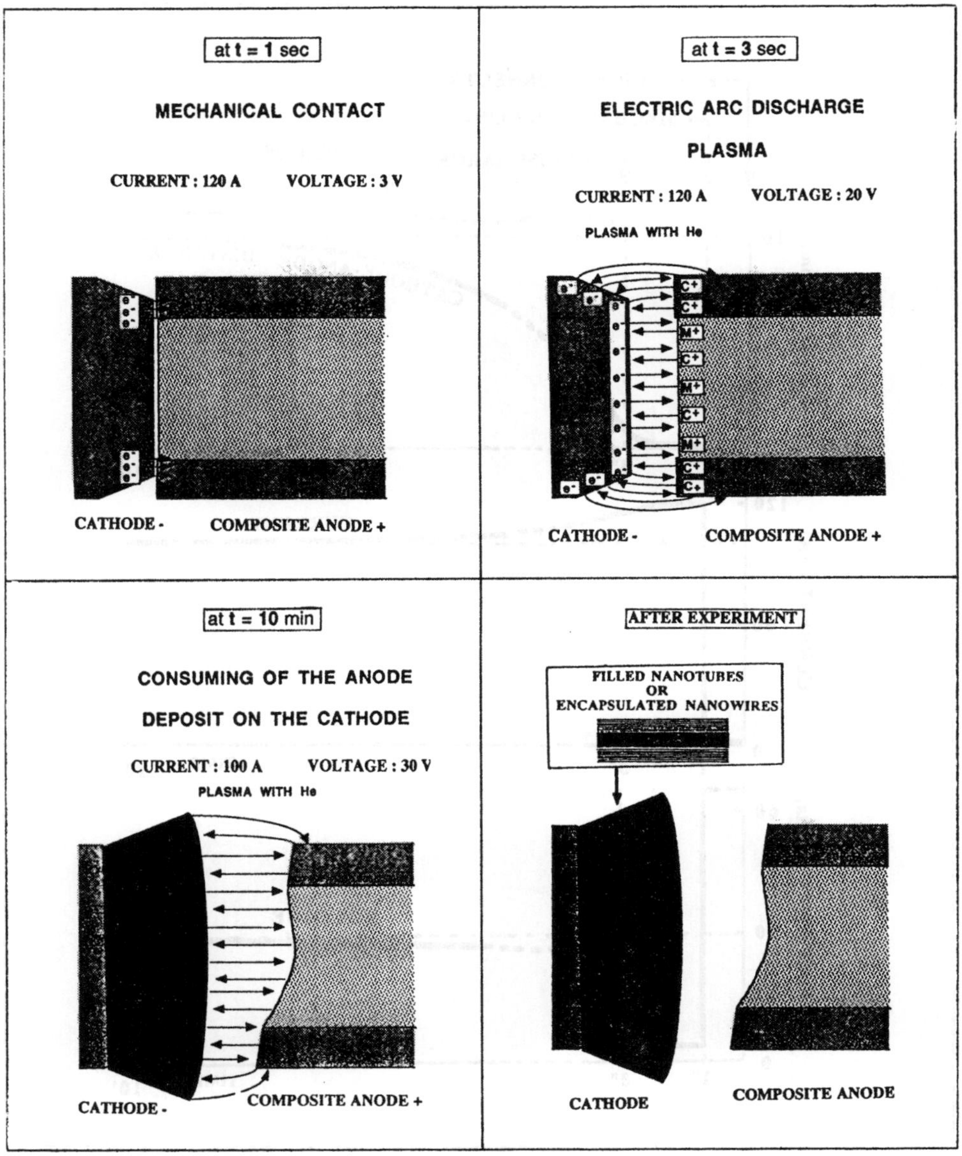

Figure 2. Schematic drawing of the different steps of one typical experiment for the synthesis of nanowires encapsulated in carbon nanotubes

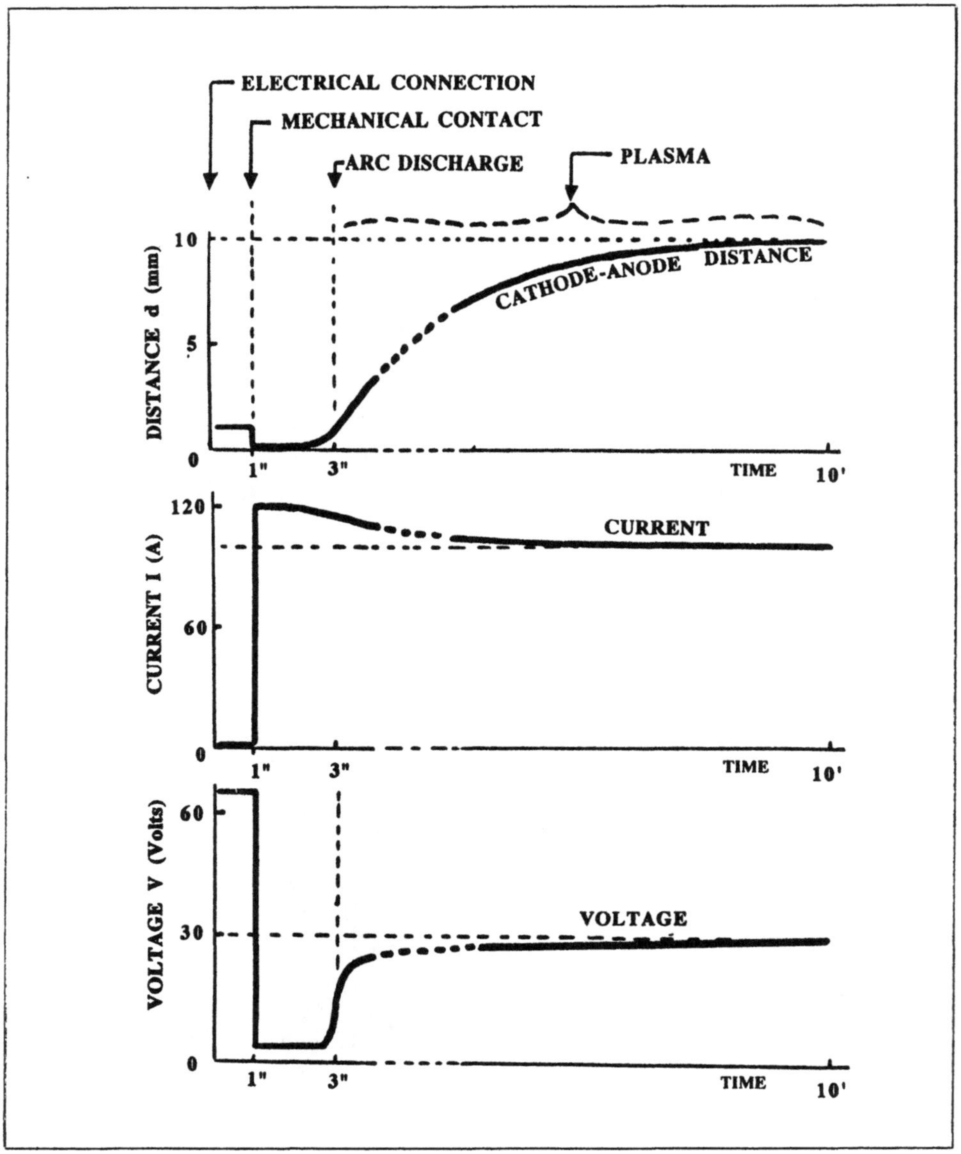

Figure 3. Evolution of the electric voltage V and electric current I, versus time, with a variable cathode-anode distance.

The place where these nanomaterials are found in the reactor, is related to the spatial dependance of the temperature during the experiment. It is known that the synthesis of fullerenes and derivates is produced in the plasma at about 5000°C. The growth of nanotubes and nanowires encapsulated in carbon nanotubes takes place mainly near the cathode surface in regions where the temperature is probably between 500°C and 3000°C.

5. Synthesized nanowires : results

Finally, we investigated the propensity to form nanowires inside graphitic nanotubes, with 43 different elements, belonging to 3d, 4d, 5d, 4f and B groups. The investigation was made by always using the same experimental conditions : 0.6 bar in helium pressure, graphite purity 99.4%, DC current 110 A, DC voltage 25 V, cathode-anode distance 5 to 10 mm, plasma duration 30 minutes. After the synthesis experiment, the deposit on the cathode was collected, ground to a fine powder, dispersed ultrasonicallly in ethanol and put on a grid. The main results are published in ref. I6I and I7I. Studies by high resolution electronic microscopy showed that long continuous nanowires encapsulated by carbon nanotubes are produced in large quantity, with 10 elements belonging to three groups of the periodic table : Ni and Cr (3d group) ; Gd, Dy, Yb and Sm (4f group) ; Ge, Se, S and Sb (B group). The main features of the synthesized nanowires are the following : crystalline nanowires are long and continuous, about 1 to 3 μ in length and 5 to 50 nm in diameter, encapsulated by 5 to 20 long and continuous graphitic nanotubes, with a composition either pure element or carbides of the element. Figure 4 shows an example of nanowires with Ni, Dy and Yb from ref. I6I.

6. Hypothesis on the growth mechanism of carbon nanowires : discussion

In the experimental procedure, the important steps are the transfer of carbon and metal ions in the plasma from anode to cathode and the deposition of these ions near the cathode surface. In these two steps, the high temperature due to the plasma and the electric field due to the electric current, probably play an important role for the nanowire growth mechanism.

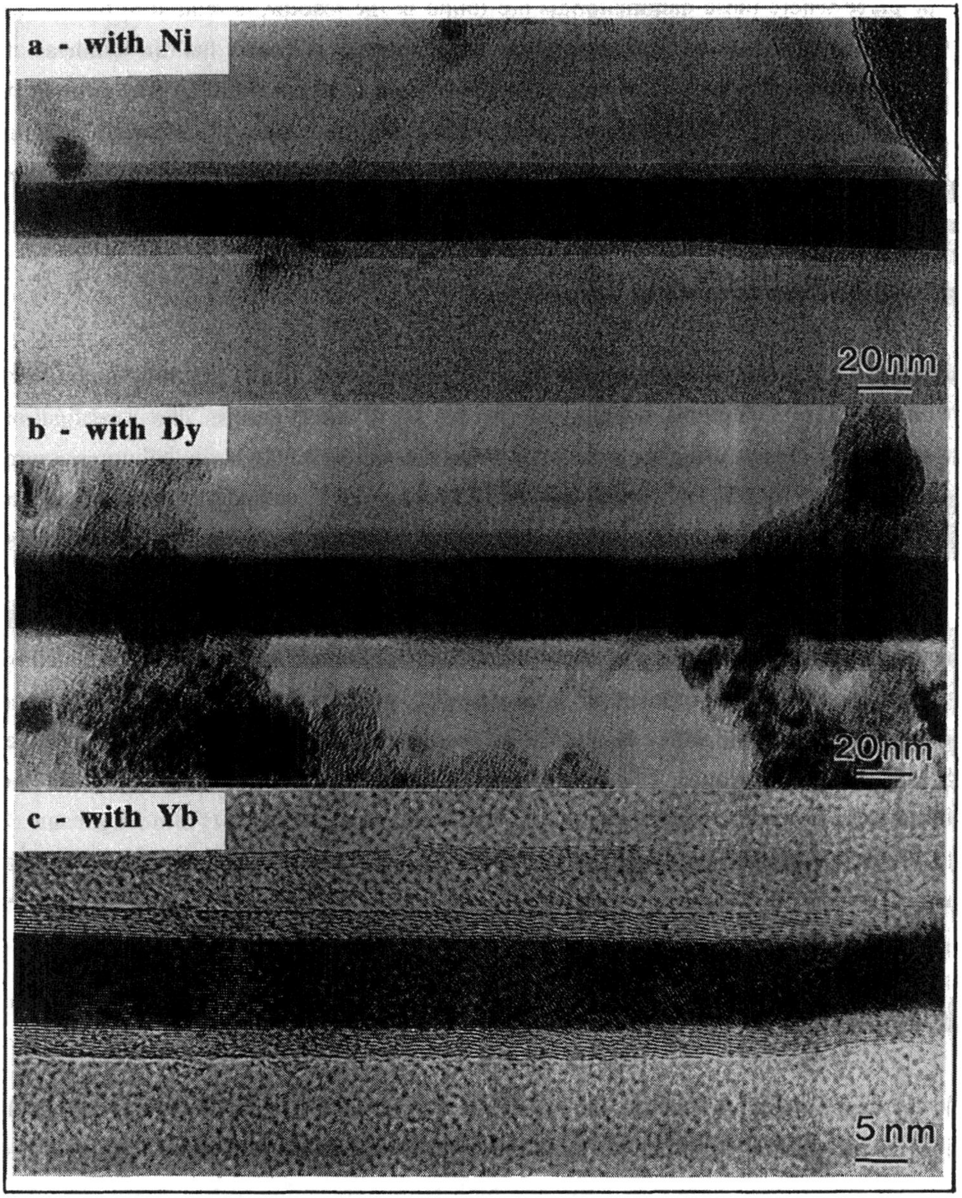

Figure 4. Examples of nanowires, with different elements, encapsulated in carbon nanotubes, from I6I.

From the ionic point of view, if we consider the electronic structure of the atoms, we observe a relation between the propensity of forming nanowires and some features of the ionic configuration. In the case of the 3d and 4f groups, this propensity seems to be related to the existence of an incomplete electronic shell in the most stable ionic state of the metal I6I. In the case of the B-group, this propensity seems to be related to the possibility of the oxidation state being positive (cationic) or negative (anionic state) I7I.

From the thermal point of view, the observations by electron microscopy show that one part of the synthesized nanomaterials on the cathode is similar to the nanoobjects grown by catalytic chemical vapor deposition, but another part is specific to the arc discharge method and particularly the long nanowires encapsulated in long carbon nanotubes. This suggests that one condition required for the filling of carbon nanotubes is the existence of very high temperatures, between 1000°C and 3000°C. Therefore, the thermal conditions (temperature ranges, temperature gradients) play a crucial role in the nanowire growth mechanism.

It seems likely that different kinds of nanowire growth mechanisms may occur, based on various chemical properties such as : catalytic reactions at high temperatures, phase transformations in liquid state or solid state, capillarity phenomena, epitaxial growth , diffusion of carbon in metals, chemical reactions between carbon and other elements. But the exact growth mechanism of long nanowires encapsulated in long carbon nanotubes remains unknown.

7. Conclusion

One important result obtained by the arc discharge method is the successful synthesis of nanowires encapsulated in carbon nanotubes.

Recently, another success has been achieved by arc discharge : the synthesis of pure boron nitride BN nanotubes : in the same reactor, a carbon-free plasma has been established between two HfB_2 electrodes in a nitrogen atmosphere and BN nanotubes have been synthesized I8I.

So, the arc discharge method, by adjusting the various parameters, can open the way to a lot of possibilities for synthesize various kinds of new nanowires.

References

1. Krätschmer, W., Lamb,L.D., Fostiropoulos, K. and Huffman D.R. (1990) Solid C60 : a new form of carbon, *Nature* **347**, 354-358.
2. Iijima, S. (1991) Helical microtubules of graphitic carbon, *Nature* **354**, 56-58.
   Ebbesen, T.W. and Ajayan, P.M. (1992) Large-scale synthesis of carbon nanotubes, *Nature* **358**, 220-222.
3. Ajayan, P.M. and Iijima, S. (1993) Capillarity-induced filling of carbon nanotubes, *Nature* **361**, 333-334.
   Ajayan, P.M., Ebbesen, T.W., Ichihashi, T., Iijima, S., Tanigaki K. and Hiura, H. (1993) Opening carbon nanotubes with oxygen and implications for filling, *Nature* **362**, 522-525.
4. Tsang, S.C., Chen, Y.K., Harris, P.J.F. and Green, M.L.H. (1994) A simple chemical method of opening and filling carbon nanotubes, *Nature* **372**, 159-162.
5. Seraphin, S., Zhou, D., Jiao, J., Withers, J.C. and Loutfy, R. (1993) Yttrium carbide in nanotubes *Nature* **362**, 503.
6. Guerret-Piécourt, C., Le Bouar, Y., Loiseau, A. and Pascard, H. (1994) Relation between metal electronic structure and morphology of metal compounds inside carbon nanotubes, *Nature* **372**, 761-765.
7. Loiseau, A. and Pascard, H. (1996) Synthesis of long carbon nanotubes filled with Se, S, Sb and Ge by the arc method, *Chem. Phys. Letters* **256**, 246-252.
8. Loiseau, A., Willaime, F., Demoncy, N., Hug, G. and Pascard, H. (1996) Boron nitride nanotubes with reduced numbers of layers synthesized by arc discharge, *Phys. Rev. Letters* **76**, 4737-4740.

MOLECULAR RECTIFICATION*

S. ROTH, S. BLUMENTRITT, M. BURGHARD, C.M. FISCHER,
C. MÜLLER-SCHWANNEKE, J. MUSTER, G. PHILIPP
Max-Planck-Institut für Festkörperforschung,
Heisenbergstr. 1, D-70569 Stuttgart, Germany

Abstract

The concept of molecular rectification is outlined and a historical survey on rectifying experiments using sandwich structures or the scanning tunneling microscope is given. Transport through Langmuir-Blodgett micro-sandwiches of a few organic monolayers between gold electrodes is discussed in terms of resonance tunneling and Coulomb charging. Asymmetric current voltage characteristics are taken as evidence for „molecular rectification". For unambiguous assignment better control of the organic/ inorganic interface is required.

1. Introduction

The classical solid state rectifyer is a p-n junction in a semiconductor. The relevant band structure is shown on the left hand side of Fig. 1, where (a) presents the flat volume and conduction band of an undoped semiconductor and (b) the band bending in a p-n junction. The bending comes about from the electrostatic interaction with the dopant ions. Simply speaking rectification reflects the fact that the electrons prefer to move „downhill" rather than „uphill". Already in the 1970'ies the question was asked whether rectifying properties could not be implemented in a single molecule. The right hand side of Fig. 1 shows the relevant molecular orbitals of a molecule with conjugated double bonds. Of course, a molecule cannot be doped in the same way as a bulk crystal but different electron affinities can be obtained by placing the conjugated moieties into different chemical environments. Figure 2 shows the chemical structure of the rectifying molecule proposed by Aviram and Ratner [1]. The conjugated moieties are tetracyanoquinone (TCNQ) and tetrathiofulvalene (TTF). They are linked by an electrically inert bridge (σ bridge). TCNQ and TTF are well known organic acceptors and donors, respectively, each of them can be crystallized separately and the stoichiometric mixed crystal TTF-TCNQ is one of the most famous organic charge transfer salts [2].

* This paper is devoted to the 65. birthday of Prof. Michael Hanack in Tübingen

C. Joachim and S. Roth (eds.), Atomic and Molecular Wires, 109-118.

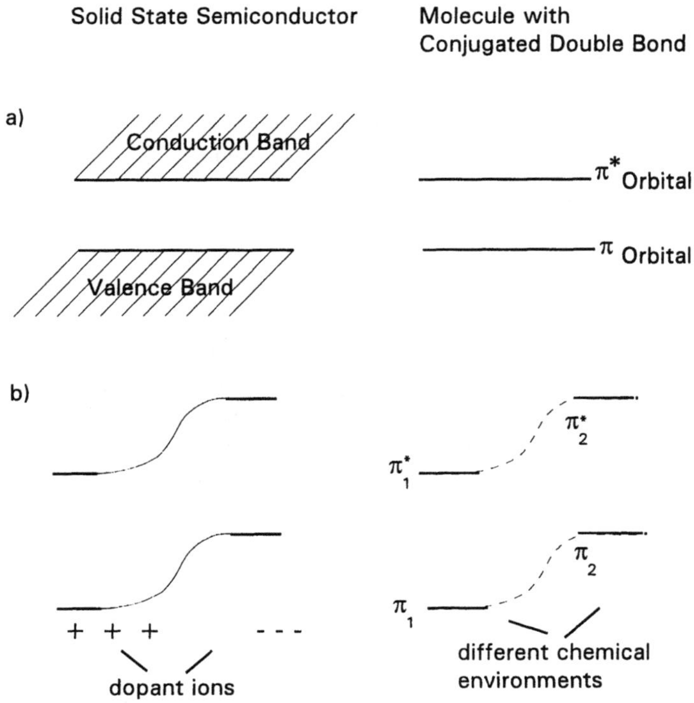

Figure 1. Comparison of semiconductor band structure and molecular levels of a π-conjugated molecule. a) undoped case, b) band and orbital shifts after doping or modification at the chemical environment.

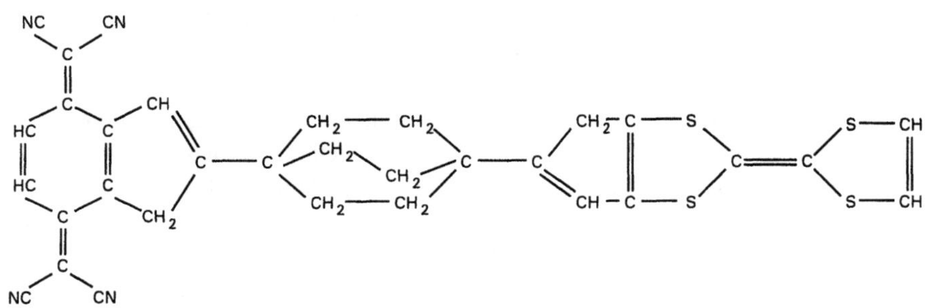

Figure 2. Rectifying „D-σ-A" molecule proposed by Aviram and Ratner [1].

In addition to molecules with σ bridges between donor and acceptor there are also molecules with π bridges. An example is shown in Fig. 3 [3]. The π bridge leads to a strong coupling between acceptor and donor and hence to a intramolecular charge transfer and a permanent dipole moment. Because this molecule has a cationic and an anionic part it is called „zwitterionic".

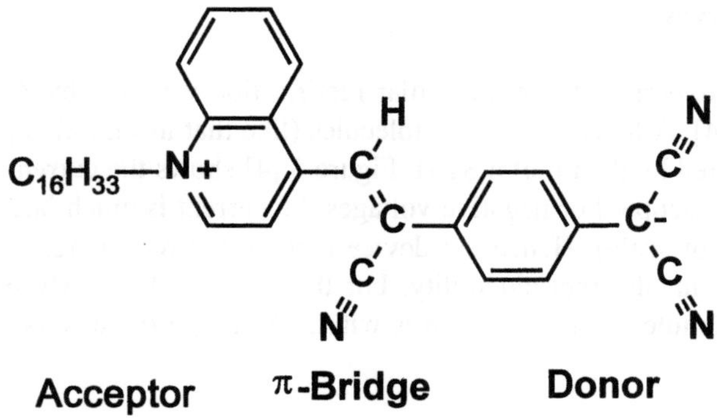

Figure 3. Zwitterionic „D-π-A" molecule used by Martin et al [3] to prepare rectifying LB layers.

In molecular rectification a third type of conjugated molecules is important. They have only one conjugated system which is relevant. These molecules are placed in a geometrically asymmetric position between metal electrodes, and the molecular level is energetically close to the Fermi level of the electrodes. The later sections of this paper will deal with this third type of molecular rectification.

A general problem is to connect a molecule to external contacts so that electrical transport and hopefully molecular rectification can be experimentally studied. Three principal ways have been proposed and they are schematized in Fig. 4: a) split electrode arrangement, b) STM tip, and c) sandwiched Langmuir-Blodgett films.

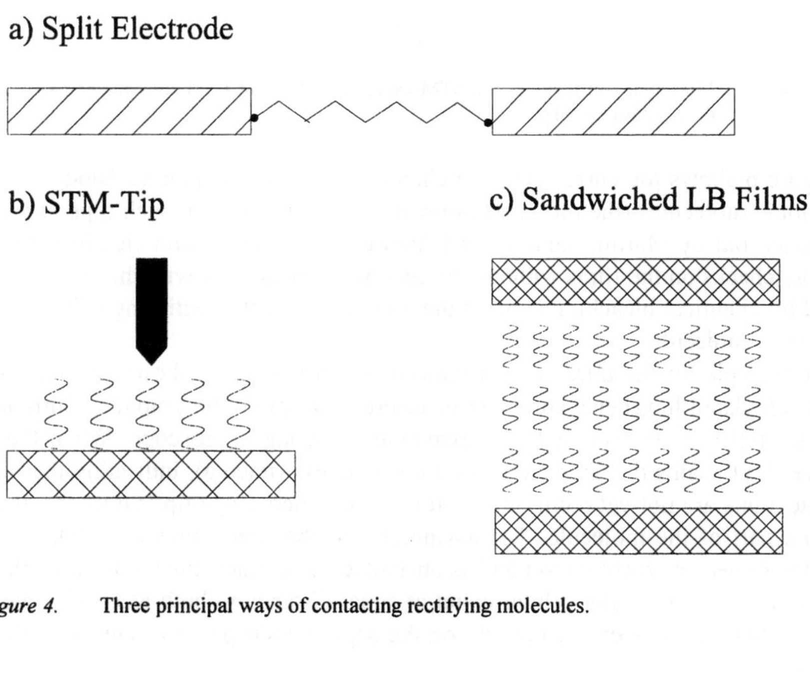

Figure 4. Three principal ways of contacting rectifying molecules.

2. Historical Survey

One of the first experiments on molecular rectification was that by Aviram, Joachim, and Pomerantz [4]. A layer of D-σ-A molecules (like that in Fig. 2) is placed on a substrate and contacted by the tip of a STM. Figure 5 [4] shows the current and the voltage trace on an oscilloscope. For negative voltages the current is much larger (in the figure it even runs out of scale). Hence the device certainly is rectifying. Admittedly, there were problems with the reproducibility, but the merit of this early experiment is to demonstrate that molecule rectification is within the range of sensitivity of our instruments.

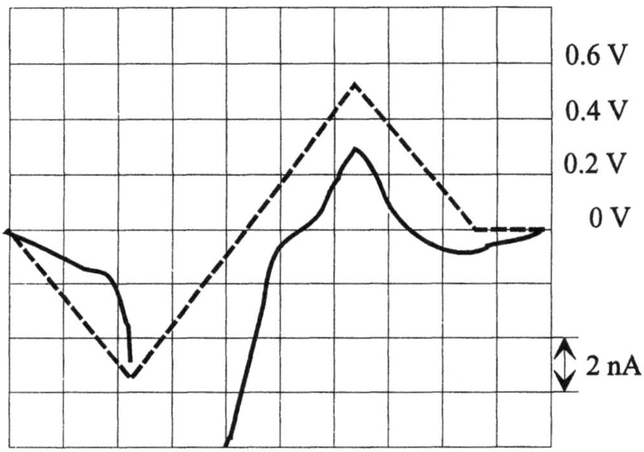

Figure 5. Molecular rectification in a STM experiment. Current (—) and voltage (----) traces on the oscilloscope [4].

Figure 6 shows the current-voltage characteristics on a Langmuir-Blodgett film of zwitterionic molecules (the molecules shown in Fig. 3). The sandwich type experiment was carried out by Martin, Sambles and Ashwell [3]. Rectification clearly is observed and cross checks on the films without zwitterionic molecules or with the zwitterions destroyed by chemical bleaching support the assignment of the rectifying behavior to the zwitterionic molecule.

An example for the third type of molecular rectification is shown in Fig. 7 [5]. A layer of alkylated hexabenzocoronene is adsorbed on pyrolytic graphite. Current-voltage characteristics are recorded by a scanning tunneling microscope. When the tip is over the alkyl chains the characteristics are symmetric (trace a), indicating that tip and substrate alone are not yet rectifying. In trace b the microscope tip is over the aromatic coronene part of the molecule. The asymmetry of the characteristics is taken as evidence for molecular rectification and is attributed to resonant tunneling through the π orbital (when the molecular orbital is not at equal distance to both electrodes the resonant condition depends on the polarity on the applied voltages as discussed in the next section).

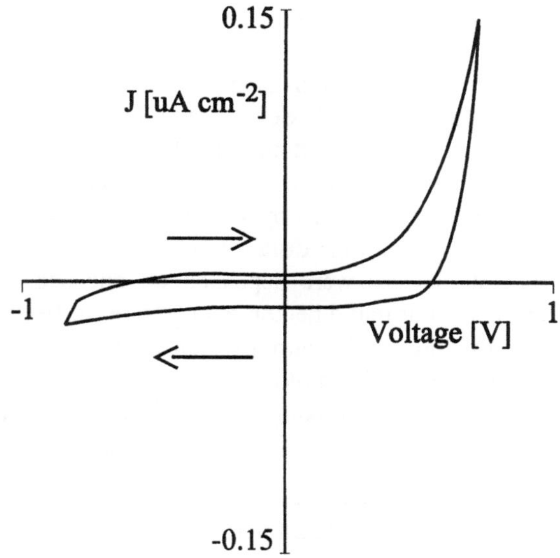

Figure 6. Current-voltage characteristic of LB films of the zwitterionic molecule in Fig. 3 (after Martin et al. [3]).

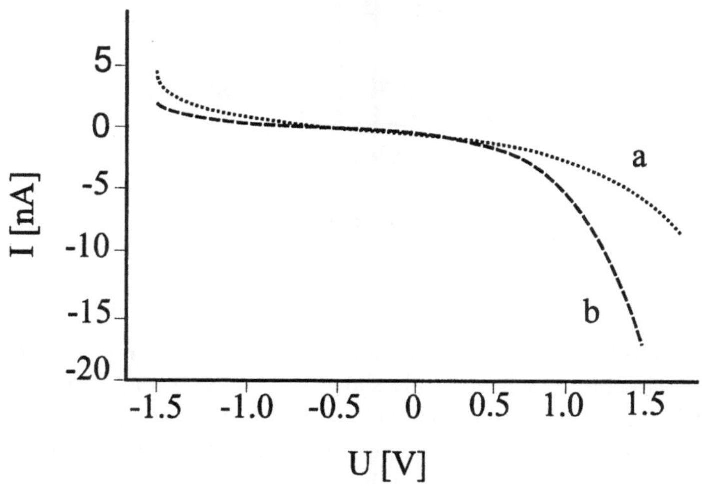

Figure 7. Tunnel characteristics through alkylated hexabenzocoronene adsorbed on pyrolytic graphite. Trace a): symmetric characteristic when tip is over alkylated part; trace b) rectifying characteristic when tip is over aromatic coronene part.

Very recently Metzger et al. [6] have observed well-pronounced rectification on LB films of Ashwell's zwitterionic molecule contacted by a STM tip. For an earlier review on molecular rectification see Metzger [7].

3. Langmuir-Blodgett Micro-Sandwiches

In our work we use sandwiched Langmuir-Blodgett films as contact configuration. A „device" is schematically shown in Fig. 8. Several monolayers of „electroactive" molecules are sandwiched between top and bottom gold electrodes. In such a configuration pinholes frequently lead to short circuits. To avoid this problem micro-sandwiches were used. The typical cross section is some 10 by 10 μm, so that there is a fair chance to find pinhole-free devices (if the average distance between two pinholes is larger than 10 μm). Usually 20 to 50 devices were deposited on one substrate and transport measurements were carried out at liquid helium temperature (to freeze-out leakage currents and thus to reveal structures in the current-voltage characteristics). In a 10 μm by 10 μm cross section there are about 10^8 molecules. In a later stage one could imagine to dilute the electroactive molecules by mixing them with inert molecules and thus one could hope to contact only a few or even individual molecules by this technique [8].

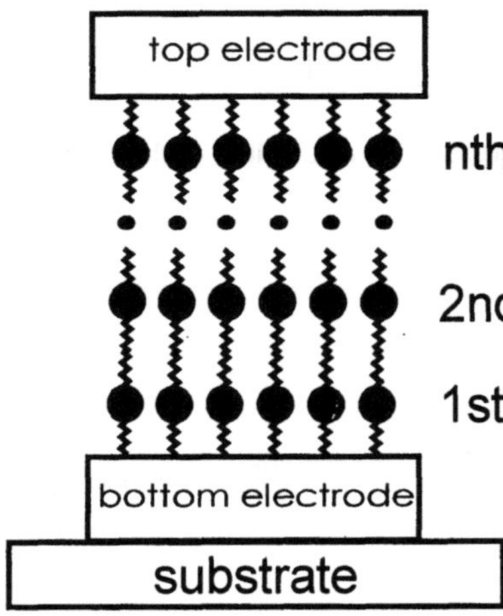

Figure 8. Langmuir-Blodgett micro-sandwich for transport studies (ML=monolayer).

As electroactive molecules we use molecules with only one π-conjugated system. The electronic structure is schematically shown in Fig. 9. In Fig. 9a the relevant π orbital (HOMO) is geometrically in the center between the gold electrodes and its energetic position is by the amount of Δ below the Fermi level of gold. The expected current voltage characteristic is indicated at the right hand side of the figure. No current is flowing until the π level gets into resonance with E_F of one of the electrodes at $V=\pm 2\Delta$. (The threshold is at 2Δ and not at Δ because the π position also changes with the applied voltage). In Fig. 9b the electroactive molecule is off-center and the characteristic is asymmetric. For $3\Delta/2<|V|<3\Delta$ the device is „rectifying". The saturation value of the current above the threshold voltage is determined by the resonant tunneling probability and the density of states at the electrodes.

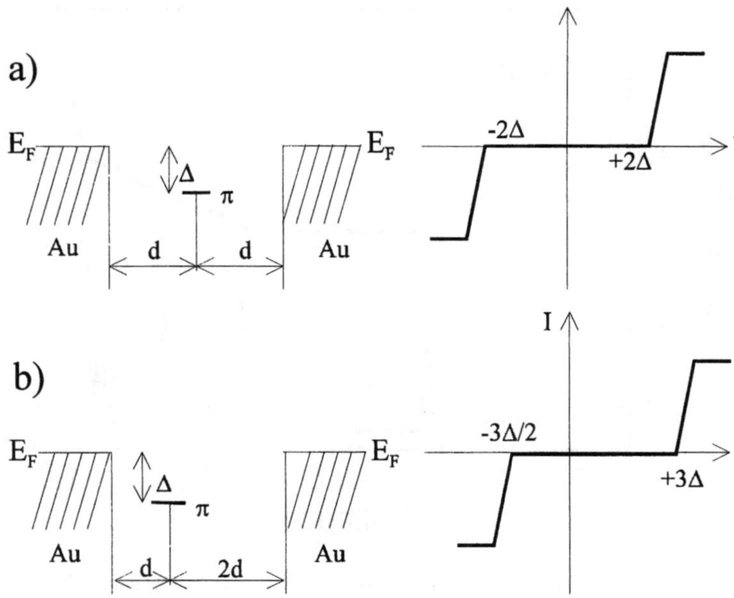

Figure 9. Energy scheme and expected current-voltage characteristics for LB sandwiches in the flat-band approximation.

This consideration is valid only for the „flat band case", i.e. if there is no spontaneous charge transfer between the electrodes and the molecule. This should be true for perfect molecules between perfect electrodes. In real devices the flat-band condition will not be fulfilled because there will be defect states at the organic/inorganic interface, so that space charges are built up and the electronic levels shift.

Figure 10 shows the current voltage characteristics of a Langmuir-Blodgett microsandwich consisting of 6 monolayers of a phthalocyanine derivative (PcPd) and 6 monolayers of a perylene derivative (PTCDI) between gold electrodes [9]. The HOMO of PTCDI is about 0.7 eV below the Fermi level of gold and that of PcPd about 0.2 eV. We assume that only the latter is relevant (because of its closeness to E_F). Because of the nonlinear dependence of the tunneling probability on distance one can easily show that the layer closest to the center position between the electrodes is dominating the transport behaviour. This layer still is off-center. So we expect current-voltage characteristics similar to Fig. 9b. And indeed, these are the main features of the experimental result in Fig. 10: asymmetric thresholds. But in addition we observe more or less regular small steps in the experiment. The inset of Fig. 10 shows also the derivative of the current-voltage characteristics and from this we see that the steps are equidistant. This behaviour is reminiscent of a „Coulomb staircase" in quantum dots, where single electrons tunnel more quickly into the dot than they tunnel out and thus the accumulated charge changes the electrostatic potential of the dot (Coulomb barrier) [10].

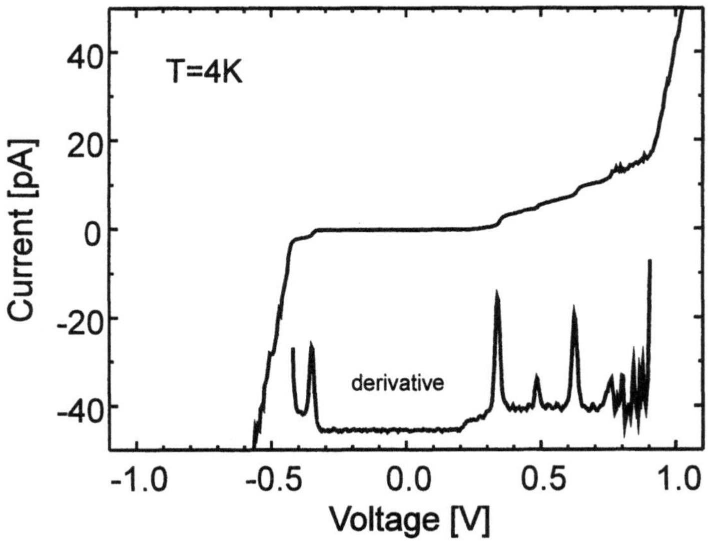

Figure 10. Rectifying current-voltage characteristic of a LB micro-sandwich consisting of 6 monolayers of a phthalocyanine derivative and 6 monolayers of a perylene derivative between gold electrodes [9].

From the voltage separation of the steps, the capacitance of the dot can be estimated and from this its diameter. The values obtained agree roughly with the diameter of a phthalocyanine molecule. However, in contrast to a semiconducting or metallic dot, the electronic structure of a molecule does not allow to accommodate several electrons at equidistant levels. We would have to assume that the dots are stacks of interacting molecules. An alternative explanation would be to evoke gold clusters which might have separated from one of the electrodes and migrated into the organic layer.

To check the gold cluster hypothesis LB sandwiches of inert layers (without π electrons) have been prepared. Figure 11 shows the current-voltage characteristic of a micro-sandwich with 4 monolayers of a cysteamine salt of PolC5F, a saturated polymer. As expected we see no thresholds – but a beautiful Coulomb staircase [11]. Originally we had hoped that the mercapto groups of cysteamine would stabilize the gold surface. Apparently they do the opposite and favour the formation of gold clusters. At this point one can argue that the curve in Fig. 10 is not a superposition of tunneling into phthalocyanine and into gold, but that only gold clusters of several sizes and inhomogeneities of the electrodes are involved. To obtain an unambiguous assignment of the features in the current-voltage characteristics more knowledge of the energetic positions of the π orbitals (with respect to the Fermi levels of the electrodes) and of their geometrical distance to the electrodes would be needed. The situation is complicated by the fact that apparently the flat-band condition is far from being fulfilled. Kelvin probe studies [12] have shown that in the first one or two monolayers of a PcPd film space charges corresponding to several hundred meV build up. Consequently further investigations will have to focus on the electronic properties of the organic/inorganic interface.

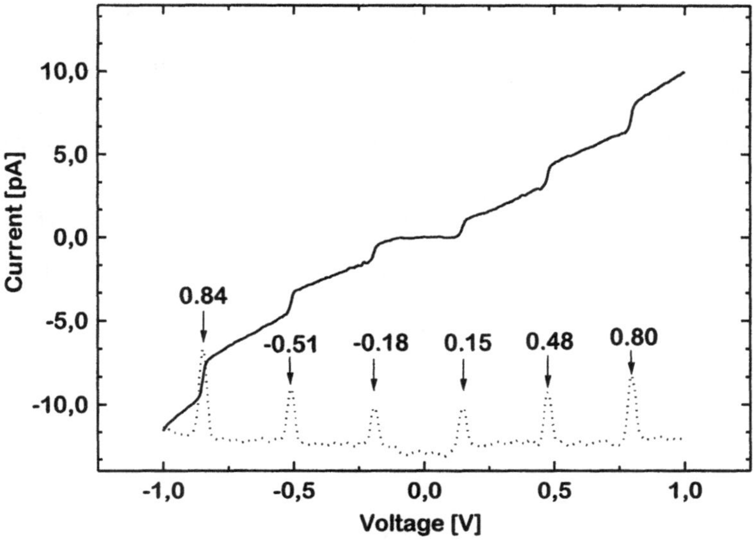

Figure 11. Current-voltage characteristic of a LB micro-sandwich with 4 monolayers of the saturated polymer PolC5F/cysteamine. The structure is interpreted as Coulomb staircase originating from a gold cluster charged by single electrons. The gold cluster might have separated from one of the electrodes and migrated into the film [11].

4. Conclusions

We have shown that short-circuit free micro-sandwiches of LB films can be prepared and that electrical transport on these sandwiches can be measured. The current-voltage characteristics exhibit distinct features like thresholds and steps, which can be interpreted as resonant tunneling into molecular orbitals and/or as single electron Coulomb charging of metallic islands (perhaps gold particles which have separated from the electrodes and migrated into the film). For an unambiguous assignment more knowledge and better control of the organic/inorganic interface is needed.

5. Acknowledgment

This work was performed in the frame of DFG Sonderforschungsbereich „Molekulare Elektronik" and EU/EMRS Network NEOME and in close cooperation with the groups of Profs. M. Hanack and W. Göpel in Tübingen.

6. References

1. Aviram, A. and Ratner, M.A. (1974), *Chem. Phys. Lett.* **29**, 277.
2. for an introduction see e.g.: Roth, S. (1995), *One-Dimensional Metals*, VCH Weinheim.

3. Martin, S., Sambles, J.R. and Ashwell, G.J. (1993), *Phys. Rev. Lett.* **70**, 218.
4. Aviram, A., Joachim, C. and Pomerantz, M. (1988), *Chem. Phys. Lett.* **146**, 490.
5. Stabel, A., Herwig, P., Müllen, K. and Rabe, J. (1995), *Angew. Chem.* **107**, 1768.
6. Metzger, R.M., Tachibana, H., Wu, X., Hoepfner, U., Chen, B., Lakshmikantham, M.V. and Cava, M.P. (1996), *Synthetic Metals* (Proceedings of ICSM'96, Snowbird, Utah, 1996), in print.
7. Metzger, R.M. (1994) in R.R. Birge (ed.), *Biomolecular Electronics*, American Chemical Society Advances in Chemistry Series **240**, p. 81.
8. Kuhn, H. (1989), *Thin Solid Films* **178**, 1.
9. Fischer, C.M., Burghard, M., Roth, S. and von Klitzing, K. (1994), *Europhys. Lett.* **28**, 129.
10. Su, B., Goldmann, V.J. and Cunningham, J.E. (1991), *Bull. Am. Chem. Soc.* **36**, 400.
11. Philipp, G., Burghard, M., Müller-Schwanneke, C., Weis, J. and Roth, S., to be published.
12. Iwamoto, M. et al., to be published.

NANOPATTERNING OF CONDUCTING POLYMERS AND CONDUCTIVITY MEASUREMENTS WITH NANOMETER-SPACED ELECTRODES

S.H. MAGNUS PERSSON,
Department of Physics,
Chalmers University of Technology and University of Göteborg,
S-412 96 Göteborg, Sweden

PETER DYREKLEV AND OLLE INGANÄS
Laboratory of Applied Physics, Department of Physics,
Linköping University,
S-581 83 Linköping, Sweden

Abstract

Direct patterning of 50 nm wide lines of the conjugated polymer poly(3-octylthiophene) is possible by exposing a thin film with a 50 kV electron beam. The exposure reduces the solubility and chlorobenzene can be used as developer to dissolve unexposed polymer. The conductivity of an exposed and developed line, consequently doped with $FeCl_3$ is of the order 1 S/cm and is comparable to that of unexposed polymer. It shows a non-linear current voltage characteristic and a temperature dependence of the conductance that indicate hopping transport and Coulomb blockade phenomena.

Lateral electrodes separated by a self-assembled monolayer (SAM) of insulating molecules is made by shadow evaporation of two gold electrodes. The molecules self-assemble on the gold surface of one electrode when the substrate is immersed in a solution of 1-octadekanethiol in hexane, before the second electrode is deposited. The resulting minimal separation is determined by the thickness of the organic monolayer, which is 1 nm. The electrodes are separated and have a resistance in the GΩ regime, but the current voltage characteristic is non-linear, with an exponentially increasing tunnelling current between the electrodes.

1. Introduction

Organic molecules have a potential to be used as the active part of electronic devices. Devices with molecular-scale dimensions and low power consumption, suitable for very large scale integration and three-dimensional architecture. Organic molecules have a well-defined structure and have a naturally small dimension that is necessary for a single-electron tunnelling (SET) device [1] to work at room temperature. Metallic and semiconducting SET-devices can be downscaled, but the lithographic limit is today tens

C. Joachim and S. Roth (eds.), Atomic and Molecular Wires, 119-128.

of nanometers, which is a slightly too large for SET-devices operating at room temperature. The number of atoms and the geometry of conducting islands made with electron-beam lithography are difficult to control and result in a variation of device properties, for example, an often chaotic distribution of energy levels, that will cause large fluctuations as dimensions shrink. The geometry and energy levels of a molecule are determined by the stoichiometry and that is the same for all identical molecules. Molecules with the same properties can be produced in very large numbers, they can be tailored to have desired electrical functionality and with an adequate choice of reactive end-groups they can self-assemble to noble metal surfaces[2,3]. Properties that can ensure reproducible device characteristics and might give opportunities to bypass the difficulties with alignment of nanometer-scale devices.

To characterise a single conducting molecule, or sufficiently few not to average out their microscopic properties, we need electrodes with a separation that is smaller than the molecule's length and a method to select only a few molecules that bridge the gap. Bumm et al. [4] have managed to measure conductance for single conjugated molecules with a scanning tunnelling microscope (STM). They isolated the conducting molecule with a self-assembled monolayer (SAM) of insulating n-dodecanethiol molecules, in which the conducting molecule was inserted in pin-holes and chemisorbed to the underlying gold surface by a thiol end-group. The STM is excellent because of its ability to image the surface and locate isolated molecules, but thermal drift and mechanical vibrations make spectroscopy difficult. An alternative is the mechanical break junction [5] that is stable, but cannot isolate nor image the molecules. Jones et al. [6] have made preliminary measurements with self-assembled 1,4-benzenedithiol in a breakjunction. However, the STM is practically impossible to use for implementing a useful electronic device and the mechanical break-junction is hard to integrate on a larger scale. It is therefore important to develop methods to contact single molecules that are both stable and possible to integrate with other electronic devices.

We have made lateral electrodes on an oxidised Si substrate that have separations comparable to the length of conducting polymers and oligomers. The electrodes are made by angle evaporation and the average distance between them can be controlled with an accuracy of a few nanometers. The roughness of the electron-beam lithography resist mask limits the minimal separation to slightly less than 10 nm before they overlap and become electrically shorted. A SAM of 1-octadekanethiol can be applied to the first electrode, by immersion of the substrate in a hexane solution. This insulating layer prevents shortage of the electrodes and the minimal separation is then determined by the thickness of the SAM, which is 1 nm. Preliminary investigations show a non-linear IV-characteristic with zero-bias resistances in the GΩ-regime.

We have developed a method for patterning poly(3-octylthiophene) (P3OT) with 50 nm resolution by exposing it to an electron beam [7]. We have used a lateral electrode structure with distances between the electrodes down to 20 nm, to measure the conductance of 50 nm wide exposed polymer strips. The polymer chains are long enough to span over the electrode gap. Hence, it is possible to investigate microscopic properties of a limited number of molecules. We observe linear current-voltage characteristics for the exposed strips, as well as for oriented polymers at high doping

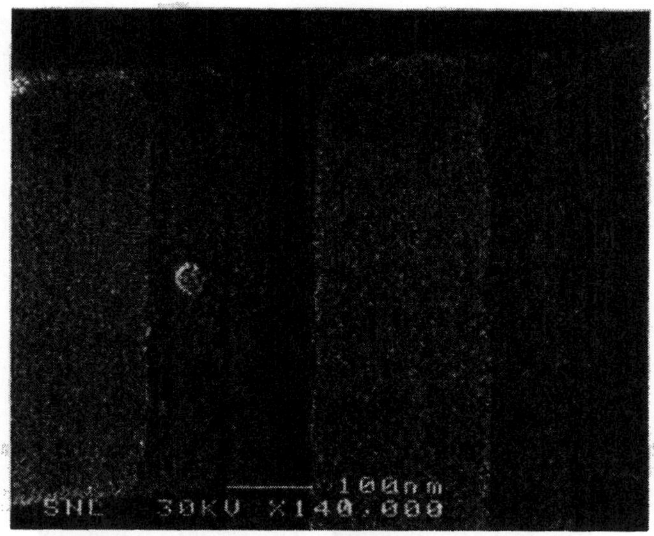

Figure 1. Gold electrodes with separation less than 10 nm made by shadow evaporation technique.

levels. At low temperature, or at low doping levels at room temperature the IV-characteristics resemble a Coulomb blockade of tunnelling, a characteristic of tunnelling between very small conducting particles.

2. Fabrication of electrodes with less than 10 nm gap

The electrode material is chosen to be gold, that give an ohmic contact to conducting polymers and a possibility for self-assembly of thiol-terminated molecules[3]. The contacts were fabricated by evaporation at two different angles relative to an e-beam defined resist mask, hanging just above the oxidised silicon substrate. We used a two-layer resist with a 170 nm copolymer P(MAA-MMA) bottom layer and a 70 nm PMMA top layer. The bottom layer has a higher sensitivity to e-beam and after development it has been dissolved to a lager extent than the top layer, resulting in a large undercut. The top resist was developed for 15-20 s in a toluene:isopropanol (1:5) mixture and the bottom resist for 8 s in 2-ethoxyethylacetate:ethanol (1:3). The substrate is tilted during metal evaporation so that the difference in position between metal deposited from two different angles can be controlled very accurately. Two shifted patterns can be seen in figure 1, leaving only a narrow gap between the electrodes. A 50 Å thick layer of NiCr alloy or Cr was deposited to improve adhesion to the substrate before the 150 Å thick gold layer. The shadow evaporation technique has been commonly used to fabricate ultrasmall tunnel-junctions [1]. The width of the electrodes is limited by how long that free-hanging PMMA strip can be without sagging. We have managed to make 2 μm wide electrodes.

The minimal gap that can be achieved with this fabrication method is limited by the roughness of the resist edge. The edges of the electrodes, seen in figure 1, have a roughness of 3-5 nm, and consequently a gap smaller than 10 nm will most probably be

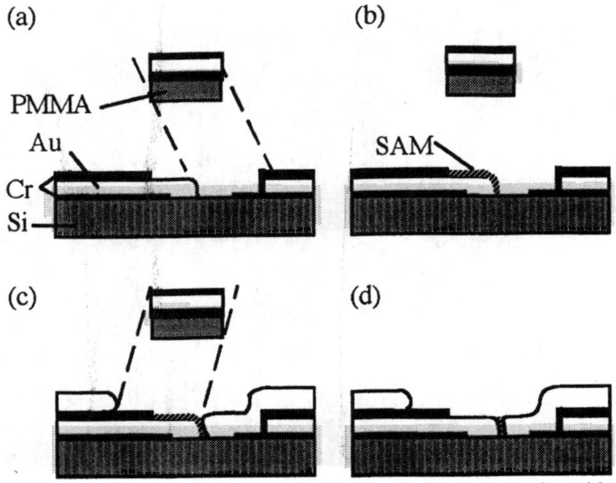

Figure 2. Schematic diagram showing the process for making isolated electrodes with small gaps. (a) The first electrode is evaporated through the resist-mask with a free-hanging pmma-bridge. The major part of the electrode surface is covered by Cr. (b) 1-alkanethiol only form a SAM where the gold is uncovered at the gap. The molecules are physisorbed at the Cr surface and can be dissolved, whereas they chemisorb on the gold surface. (c) The second metal layer is evaporated from a different angle, as close as possible to the first electrode. (d) The structure after Lift-off and O_2 -plasma etching that remove the organic molecules, except in the narrow gap.

shortened. This problem will not be improved to any larger extent by choosing another resist or changing any other parameter of the lithography process. The n-alkanethiols are known to form well-ordered monolayers on gold surfaces[3], and to overcome the risk of shortage we protect the first electrode with a SAM of 1-octadekanethiol, an insulating monolayer that is approximately 1 nm thick. This allows us to decrease the electrode gap further, below the previous limits of the lithography process.

The process with depositions of metal and 1-octadecanethiol is shown schematically in figure 2. The substrate with the copolymer-PMMA resist mask is immersed for 30 s in 1-octadekanethiol dissolved in hexane, after the first deposition of metal. The hexane does not dissolve the resist mask. 1-alkanethiols and other thiol-terminated molecules form SAM on gold surfaces and other noble metals. A SAM covering the whole evaporated gold surface would therefore isolate it from next overlapping layer, resulting in a large tunnel junction instead of a metallic contact. An evaporation of chromium is therefore made with an angle normal to the substrate, that covers the mayor part of the gold surface except the region closest to the gap. The molecules will chemisorb to the gold, but only physisorb on Cr and they can therefore be removed. The substrate is rinsed in hexane and the physisorbed molecules will be dissolved to such extent that there will be a good electrical contact between the two layers of evaporated metal and an insulating SAM only on the gold surface.

The scanning electron microscope cannot resolve a the a gap that is less than 10 nm (see Fig. 1) and it cannot be determined if the electrodes are shortened or not. The current voltage characteristics of the nano-gaps indicate a shortage for roughly half the samples with resistances of a few kΩ. The rest have non-linear current-voltage

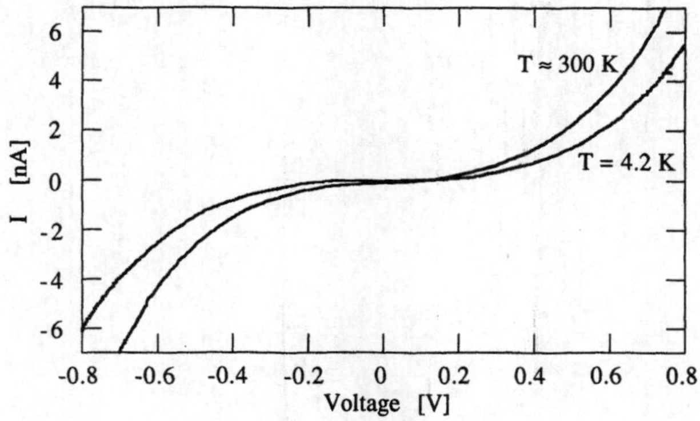

Figure 3. The current voltage characteristic of electrodes with less than 10 nm gap, after an oxygen plasma etch. Only a slight decrease in conductance was observed as the temperature was lowered from room temperature to 4.2 K.

characteristics (see Fig. 3) with zero-bias resistance in the GΩ regime after a cleaning in O_2-plasma. The two curves in Fig. 3 shows that the temperature was weak and that conductance only decrease slightly as the temperature is lowered from room temperature to 4.2 K. We can therefore exclude leakage currents on the substrate as the conduction mechanism. Only a few samples have been produced and the electrical measurements are not reproduced, therefore we only conclude that it is possible to decrease the electrode distance to less than 10 nm, without having a shortage and that the distance is short enough for a tunnelling current to flow between the electrodes. We have tried to make a SAM of conjugated molecules on the gap, but these experiments have so far not resulted in any decrease of the resistance, nor any feature that can be attributed to the conjugated molecules.

3. Nanopatterning of poly(3-octylthiophene)

The first synthesised conducting polymers were insoluble and difficult to handle, but recent developments in chemical synthesis have resulted in a number of derivatives that are soluble and stable in air. These have been used for fabrication of light emitting diodes, transistors and other macroscopic devices, where the choice of side chain determines the properties of the component. P3OT is soluble in non-polar organic solvents, due to its alkyl side chains and thin films can easily be prepared by spin coating.

A lithography process was developed to directly pattern the conducting polymer strips over the electrode gap [7]. The solubility of the undoped polymer decreases after exposure with high-energy electrons and therefore P3OT can be used as a negative e-beam resist. This simple process consists of four steps: spinning, exposure, development and doping. P3OT, dissolved in chlorobenzene (1.8% by weight), was spin coated (3000 rpm) onto an oxidised silicon substrate cleaned in acetone and isopropanol. The thickness of this polymer film was 30 nm. The substrate was baked on a hot-plate at

Figure 4. 50 nm wide strips of poly(3-octylthiophene), with a pitch between the lines of 150, 140 and 130 nm, from the left. The polymer strips are drawn together when the pitch is less than 150 nm

90°C for 2 minutes. The polymer films were exposed by an electron-beam with 50 keV acceleration voltage and 10 nm beam-diameter. The exposed dose ranged from 0.5 to 2 nC/cm for line-exposures and from 50 to 800 $\mu C/cm^2$ for area-exposures. The polymer was developed by soaking in chlorobenzene for 10 s, then rinsed in isopropanol and blown dry with nitrogen. A minimum line width of 50 nm was accomplished. Figure 4 shows polymer strips of conducting polymer made by line exposures. The linewidth is 50 nm and the pitch is 150, 140 and 130 nm, respectively. The resolution limit is reached at a pitch of 140 nm, where the different strips are linked and form an irregular network. We estimate the number of polymer chains in a crossection of one strip to be less than thousand.

We used Fourier-transform infrared spectroscopy of the polymer at various doses of e-beam exposure to ascertain possible degradation during the irradiation. The polymer was spin coated on a silicon wafer with low doping, giving transparency in the IR range. Experimental studies were done in transmission. With the low signal intensity from a thin film prepared after e-beam treatment, less than 120 nm, it was not possible to observe any new vibrations, nor to comment on minor changes of any of the major peaks. It proves that the major parts of the conjugated polymer are still present after e-beam exposure. These studies do not indicate a major change of the chemical structure and thus only a small fraction of the chains can be crosslinked. It is known that the radiation endurance of organic polymers increases with the amount of conjugated segments; there are processes to dissipate the energy delivered to the conjugated system more effectively than in the non-conjugated polymers [8]. Cross-linking reduces the solubility of the exposed polymer and makes patterning by electron beam lithography possible. The crosslinking reactions due to irradiation of light of poly-alkylthiophenes are described by Abdou et al. [9,10] , and was also used for lithography. The resolution of their optical lithography process is limited to 1 µm linewidths.

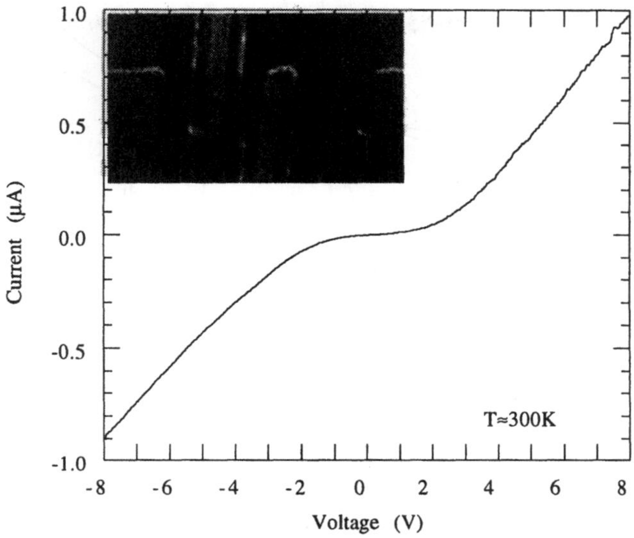

Figure 5. A 2-probe measurement of a 50 nm wide strip of P3OT doped with $FeCl_3$. The electrode gap was 20 nm. The inset shows the electrode geometry with a polymer strip bridging the gap.

4. Electrical transport measurements

We have measured the electrical properties of a number of samples with electrode gap ranging from 100 nm down to 20 nm and widths from 0.5-4 μm. The inset of Fig. 5 shows a four-probe electrode geometry where the separation is 20 nm with a patterned P3OT strip bridging the gap. The same configuration with different widths is also used for measurements of oriented P3OT films. These films where made by spinning of P3OT on a polyethylene substrate and then stretched in nitrogen atmosphere at 85 °C, at a speed of 0.25 mm/s, to 7 times its initial length. The oriented film is transferred to the four-probe electrodes with the same technique as used by P. Dyreklev et al. [11], with the molecules oriented to bridge the electrode gap.

Undoped poly(3-octylthiophene) has conductivities typically of the order 10^{-5} S/cm. The conductivity increases up to six orders of magnitude by doping with iodine or $FeCl_3$. Our substrates were dipped into a 0.1M $FeCl_3$ nitromethane solution for typically 20 s. The P3OT becomes insoluble after doping. Patterning of the polymer must therefore be done before doping. Narrow strips of highly doped, conjugated polymer gave linear current-voltage characteristic, with conductivities of the order of 1 S/cm. This is somewhat lower than, but comparable to, previous reports of the macroscopic conductivity of P3OT in the doped state [12]. Non-linear IV-curves where found at lower doping levels, shown in figure 6. The lower doping is accomplished with lower concentration of $FeCl_3$ or by waiting a few days after doping before measurements.

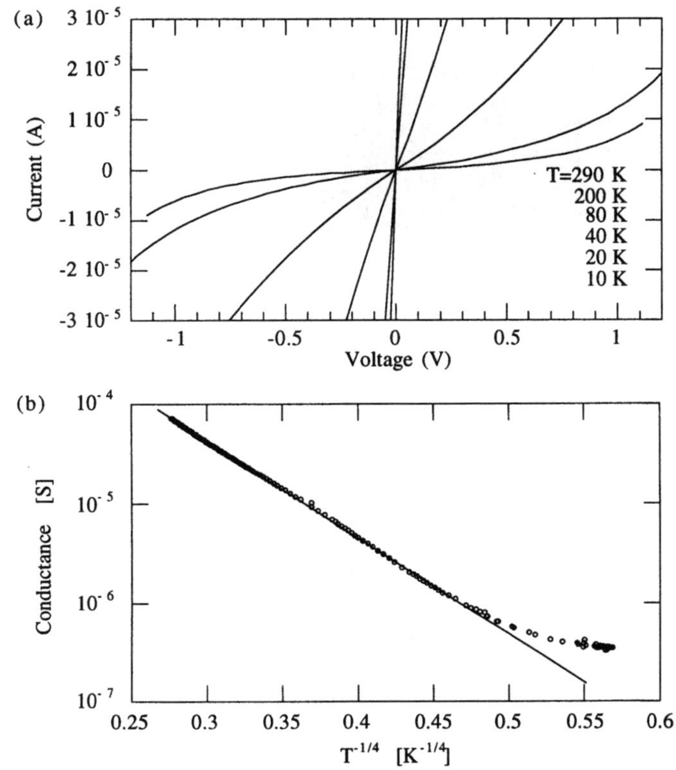

Figure 6(a). The current voltage characteristics of P3OT stretched to 7 times its initial length. (b) The zero bias conductance fitted to the straight line that indicate the theoretical variable range hopping law. A deviation is seen below 25 K.

The characteristic of figure 5 resembles a Coulomb blockade of tunnelling [3,14] that occur when single electrons tunnel to a small conducting particle through high resistance junctions, where the charging energy of a single excess electron on the particle is larger than the thermal energy. A non-linearity is also seen when the four-probe configuration is used for the oriented films where the number of molecules is considerable larger and the electrode gap is 50 nm. These larger samples show the same behaviour as the narrow strip, but the observed blockade of the current is more pronounced for the strip.

5. Discussion

The conductance of the P3OT, both the narrow strips and the wider oriented films, show a strong temperature dependence. The room temperature conductivity of undoped or lightly doped polymer is low and the charge carrier transport is described as occurring via hopping between localised polaronic or bipolaronic states[13]. The macroscopic conductance is determined by the tunnelling probability between neighbouring highly conductive regions. The conductivity will decrease to zero at low temperatures and small driving voltages, both in the case of hopping conduction and for a system where charging effects blocks tunnelling between small conducting particles. The temperature dependence of the zero bias conductance is fitted to the theory of variable range hopping[14] in a three dimensional system (see Fig. 6(b)),

$$\sigma = \sigma_2(T)\exp\left[-(T_3/T)^{1/4}\right],$$

where T_3 is a constant dependant on the density of states in the material and the prefactor $\sigma_2(T)$ is more slowly varying than the exponential factor, but there is a there is a deviation below 30 K, where also a non-linearity is seen in the IV-characteristics (see Fig 5(b)). The non-linearity resemble a Coulomb blockade of tunnelling and in this case the conducting islands would be localised charges distributed along the polymer chain at the location of dopant ions. These measurements do not show any remarkable difference compared to measurements performed with macroscopic electrodes and we cannot exclude that electrons are hopping between different chains, as in a macroscopic sample.

Measurements with conducting molecules in the sub 10 nm gap have not yet been performed. With such small gap it more probable that the conductance through a single molecule can be measured. Once we have managed to contact a single molecule and determined that it is the right molecule that is bridging the gap and not any impurity, we still need an external signal to modulate the electron transport through the molecule. It is more likely that a gate electrode can be made with the lateral electrode geometry described in this article, than with a sandwich structure[15] or with mechanical break junctions that has been used for contacting organic molecules, but how to make the third terminal and realising a molecular transistor is a problem that remains to be solved.

We have shown that it is possible to directly pattern a conducting polymer by electron beam lithography with a resolution down to 50 nm line widths. The conductivity of the doped polymer after processing was measured using the nanoprobe and was found to be comparable to that of the unpatterned polymer in the doped state, and evaluated with a macroscopic probe. The ability to accurately pattern the conducting polymer and to make very closely separated contact electrodes, with distances down to 1 nm, gives us the opportunity to measure electron transport properties of these small strips with a limited number of conducting polymer chains. This new ability, to make a separation that is equal to length of a molecule opens new possibilities in the field of molecular electronics. We may now realise a molecular-scale device, possibly based on single electron effects, that operate at room temperature.

6. Acknowledgements

We acknowledge Mats Andersson and Neste Oy. for synthesis of the conducting polymers. We used the Swedish Nanometer Laboratory for electron beam lithography. This work was supported by the Swedish Board of Technical and Industrial Development (NUTEK) through the Micronics program and the Swedish Research Council for Engineering Sciences.

7. References

1. See references in (1992) Grabert, H. and Devoret, M. (eds.),*Single Charge Tunnelling, Coulomb Blockade Phenomena in Nanostructures*, Plenum, New York.

2. Tour, J.M., Jones II, L., Pearson, L. et al. (1995) Self-assembled monolayers and multilayers of conjugated thiols, alpha,omega-dithiols, and Thioacethyl-containing adsorbates. Understanding attachments between potential molecular wires and gold surfaces, *J. Am. Chem. Soc.*, **117**, 2529.

3. Poirier, G.E. and Pylant, E.D. (1996) The self-assembly mechanism of alcanethiols on Au(111), *Science*, **272**, 1145.

4. Bumm, L.A., Arnold, J.J., Cygan, M.T. et al. (1996) Are single molecules conducting?, *Science*, **271**, 1705.

5. Muller, C.J. and Reed, M.A. (1996) There is plenty of room between two atom contacts, *Science*, **272**, 1901.

6. Jones II, L., Pearson, L., Lamba, J.J.S., Tour, J.M., Whitesides, G.M., Muller, C.J. and Reed, M.A. (1995) Conjugated Oligomers, their assembly on gold, and their uses in molecular electronics studies, *Polym. Prep.*, **36**, 564.

7. Persson, S.H.M., Dyreklev, P. and Inganäs, O. (1996) Patterning of poly(3-octylthiophene) conducting polymer films by electron-beam exposure, *Adv. Mater.*, **8**, 405-408.

8. Grubb, D.T. (1974) *J. Mater. Sci.*, **7**, 43.

9. Abdou, M.S.A., Xie, Z.W., Leung, A.M. and Holdcroft, S. (1992) Laser, direct-write microlithography of soluble polythiophenes, *Synt. Met.*, **52**, 159.

10. Abdou, M.S.A., Xie, Z.W., Lowe, J. and Holdcroft, S. (1994) Microlithography of pi-conjugated polymers, *Proc. SPIE, Int. Soc. for Opt. Eng.*, **2195**, 756.

11. Dyreklev, P., Berggren, M., Inganäs, O., Andersson, M.R., Wennerström, O. and Hjertberg, T. (1995) Polarized Electroluminecence from an Oriented Substituted Polythiophene in a Light Emmiting Diode, *Adv. Mater.*, **7**, 43.

12. Gustafsson, G., Inganäs, O., Salaneck, W.R. et al. (1991) in Brédas, J.L. and Silbey, R.(eds.),*Conjugated Polymers*, Kluwer Academic Publishers, Dordrecht, pp. 315-362

13. Stubb, H., Punkka, E. and Paloheimo, J. (1993) Electronic and optical properties of conducting polymer thin films, *Mat. Sci. Eng.*, **10**, 85-140.

14. Mott, N.F. and Davis, E.A. (1971) (eds.),*Electronic Processes in Non-Crystalline Materials*, Clarendon Press, Oxford.

15. Fischer, C.M., Burgghard, M., Roth, S. and v. Klitzing, K. (1994) Organic Quantum Wells: Molecular Rectification and Single Electron Tunnelling, *Europhys. Lett.*, **28**, 129.

TOWARDS ELECTRICAL TRANSPORT ON SINGLE MOLECULES
first results on nanofabrication and phthalocyanine polymers

C. DEKKER,* S. J. TANS, L. J. GEERLIGS, and A. BEZRYADIN

Delft University of Technology
Department of Applied Physics and DIMES
Lorentzweg 1, 2628 CJ Delft
The Netherlands

J. WU and G. WEGNER

Max-Planck-Institüt fur Polymerforschung
P.O. Box 3148, W-55021 Mainz
Germany

Abstract

We report on several steps towards experiments aimed at measuring electrical transport through a single molecular wire between two fabricated electrodes. Deposition of molecular wires of phthalocyaninepolysiloxane is described. It appears to be possible to deposit a single wire bridging two closely spaced electrodes, and to image it by high-resolution atomic-force microscopy. Transport measurements on monolayers of phthalocyaninepolysiloxane are reported. Data are consistent with the space-charge-limited-current model. In the undoped state, the conductivity appears to be too low to yield a detectable current through a single molecular wire. Electrodes are made by electron-beam lithography down to a scale of ~20 nm. Much smaller electrodes (contact spacings down to 2.5 nm) have been fabricated by electron-beam induced deposition.

1. Introduction

It is over two decades ago that the possibility of electronic circuits at the scale of single molecules was proposed.[1] The field of molecular electronics has grown considerably since then.[2] Molecular wires and switches were devised and synthesized, intricate architectures for signal processing on the basis of molecules were proposed, nonlinear-optical and magnetic effects in molecules were explored, thin-film organic FETs and LEDs were succesfully developed, et cetera. While much has been

C. Joachim and S. Roth (eds.), Atomic and Molecular Wires, 129-138.

achieved, the feasibility of the original proposal, viz., electronics at the true molecular scale, is still very unclear. A crucial test experiment would be to measure the conduction through a single molecular wire. Such experiments have not been carried out successfully yet.

Measuring electrical transport through a single molecular wire is a challenging enterprise for various reasons. One major difficulty is the interfacing to the macroscopic world. Here, the issue is the very different length scales of molecules and fabricated electrode structures. While the size of a molecule is measured in Ångstroms, or in nanometers for favorable cases, even the most advanced electron-beam-lithographic patterning of metal electrodes involves a length scale that is at least one order of magnitude larger. Other relevant issues are the deposition of molecules between the electrodes, immobilization of the molecules, and imaging of the resulting configuration of molecular wires in the gap. Next to these practical questions, more fundamental issues are unsettled as well. For example, how does one generate carriers at the single-wire scale (doping, field effect, light,...)? How is charge injected from the metal electrode into the wire? What is the mechanism for electrical conduction through a molecular wire (tunneling vs carrier transport)? How does charge transport proceed within the wire (soliton/polaron hopping vs band transport,...)? It is these fundamental questions that constitute the motivation for pursuing an experimental study of electrical transport through the basic elements of molecular electronics, starting with simple molecular wires.

Here we report the status of our efforts to study electrical transport through a single molecular wire. Our approach is twofold. First, we use "conventional" electron-beam nanofabrication of metal electrodes to study long (~100 nm) macromolecules of a phthalocyanine polymer. Single molecular wires between electrodes have been obtained (Sec. 2). The low carrier concentration in these intrinsic-semiconducting wires has up till now prevented measuring electrical transport at the single-wire level. Second, we aim at electrical transport through well-defined conjugated oligomers which have a typical size of, say, 5 nm. For this, we pursue new routes for nanofabrication of very closely spaced electrodes. Gaps down to a few nm have indeed been obtained (Sec. 3).

2. Transport through phthalocyanine polymers

We have studied phthalocyaninepolysiloxane (PCPS) polymers (Fig. 1).[3] These are 20 to 50 nm long stacks of cofacial Si-phthalocyanine monomers linked by oxygen atoms. Transmission-electron microscopy (TEM) has shown that the PCPS polymers arrange in a head-to-tail fashion, which results in molecular wires with a length of hundreds of nms.[4] Including the flexible hydrocarbon side chains, which ensure

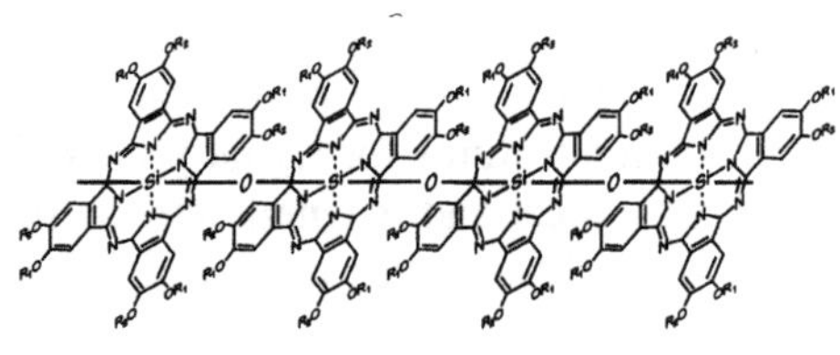

Figure 1. **Phthalocyaninepolysiloxane (PCPS) polymer. R_1=CH_3, R_2=$(CH_2)_7CH_3$.**

solubility, the total diameter of the polymer amounts to 2.3 nm. The distance between monomers is 0.33 nm. This small distance accounts for the mechanical rigidity of PCPS as well as for π-orbital overlap which renders the polymer a one-dimensional semiconductor.[5]

In Sec. 2.1 we describe the deposition and imaging of single PCPS molecular wires on fabricated electrode structures. An investigation of the nonlinear conductivity of monolayers of PCPS is reported in Sec. 2.2.

2.1. AFM IMAGING OF PCPS MOLECULAR WIRES

Individual PCPS molecules can be deposited in a controlled way in a monolayer of isopentylcellulose. By embedding the conducting PCPS polymer molecules in a monolayer of the insulating cellulose, they are strongly immobilized on the surface as well as electrically separated. The mixed monolayer of PCPS and cellulose is deposited with the Langmuir-Blodgett (LB) technique in a single dip.[4] Atomic force microscopy (AFM) is used for imaging of the molecules.[6] In order to reduce tip-sample forces, the ambient AFM (DI Nanoscope III) is operated in tapping mode (TM). In TM, the cantilever is vibrating at resonance with large amplitude (~100 nm). Changes in the amplitude resulting from the contact between tip and sample are used as the feedback signal. With this method, tip-sample friction forces are greatly reduced as compared to standard contact-mode AFM.

Figure 2a is an AFM image of a PCPS/cellulose monolayer. The PCPS polymers are organized head-to-tail which accounts for the observed length of hundreds of nms. The polymers tend to be directed along the LB dipping direction. The measured height

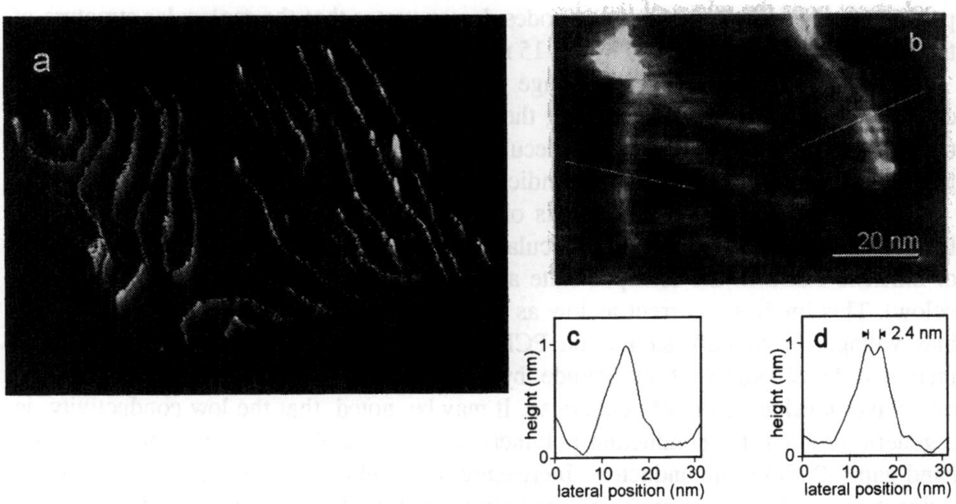

Figure 2. a. Low-resolution tapping-mode AFM image of a mixed PCPS / cellulose monolayer. Long PCPS molecular wires are observed. Image size is 700x700 nm^2.
b. High-resolution tapping-mode AFM picture of a single (left) and 2 adjacent (right) PCPS wires.
c. and d. denote the respective height profiles of these wires.

difference between PCPS and cellulose varies from 0.5 to 1.4 nm. The maximum value of 1.4 nm is in agreement with the difference of the diameters of PCPS and cellulose which amount to 2.3 nm and 0.9 nm respectively. Even in air, imaging is very stable and nondestructive. This is attributed to the immobilization of the embedded PCPS molecules and the use of the TM technique. Contact-mode AFM yielded a drastically worse imaging quality.

The AFM image of Fig. 2b indicates that it is possible to distinguish between an isolated PCPS molecule (left side of the figure) and a cluster of two directly adjacent molecules (right side). One can even discern the individual polymers within the cluster. From data of various clusters, we find an average distance between adjacent polymers of 2.4 ± 0.2 nm, in agreement with the value of 2.3 nm found from TEM and X-ray diffraction.[4] From the sharp features in the cross sections (Figs. 2c and 2d) one can infer a lateral resolution of better than 1 nm. The intra-cluster resolution shown here can only be attained with very sharp tips and for limited time (about 1 hour), which is probably due to tip wear. A problem in imaging small molecules is tip convolution, i.e., the apparent width of the AFM image of a molecule reflects the tip size rather than the intrinsic molecular width. The height profile of the isolated molecule (Fig. 2c) illustrates this effect: Its base, ~15 nm wide, is much broader than the actual polymer width of 2.3 nm. Deconvolution leads to an estimate of the tip radius of ~7.5 nm.

Electrodes are fabricated by use of electron-beam lithography on a double layer of PMMA-based resist. Through the resist mask, a 5 nm Ti adhesion layer and 10 nm of Pt are evaporated. The spacing between neighbouring electrodes is ~20 nm. As can be seen from Fig. 3, single PCPS wires can still be discerned on the Pt electrodes despite the ~1 nm corrugation of the bare Pt surface. Inspection of the pattern of the PCPS polymers near the edges of the electrodes demonstrates that the molecular structure of the monolayer is not perturbed by the 15 nm thick electrodes.[6]

It is possible to deposit and image a single PCPS molecular wire bridging two closely spaced electrodes. In covering the electrodes with a PCPS/cellulose monolayer there is a chance that an isolated molecule - or a few - bridges two electrodes. This in fact is the case with the PCPS chain indicated by the two arrows in Fig. 3.

Electrical transport measurements on such samples have shown that the current through a single undoped PCPS molecular wire is too small to detect. The conductivity of intrinsic PCPS in the undoped state appears to be of the order of 10^{-10} S/cm (see below). This implies a current as low as 10^{-18} A for a single PCPS polymer at 1 Volt bias voltage. From bulk studies on PCPS it is know that the conductivity may be increased by 9 orders of magnitude by chemical doping.[5] Doping studies at the monolayer level are currently underway. It may be noted that the low conductivity is a generic problem for conducting polymers and other molecular wires that are wide-band gap (~2 eV) semiconductors. Increasing the conduction to a reasonable value for a single molecular wire is an unexplored open question that needs to be addressed.

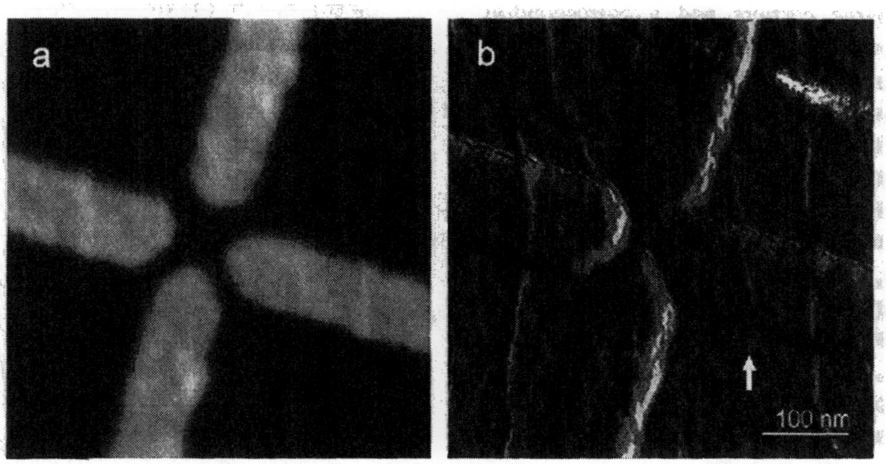

Figure 3. Tapping-mode AFM height (a) and amplitude (b) image of a PCPS/cellulose monolayer on top of an electrode structure with four leads. The height image (a) displays the 15 nm high electrodes, whereas the PCPS wires are most clearly observed in the amplitude image (b). The electrodes appear not to perturb the PCPS pattern in the monolayer. The PCPS polymer chain indicated by the two arrows bridges two electrodes.

2.2. TRANSPORT IN PCPS MONOLAYERS

We have investigated electrical transport within monolayers of PCPS.[7] Layers were transferred on interdigitated electrodes defined by e-beam lithography on a Si/SiO_2 substrate. The electrode spacing is 0.2 to 1 μm and the total effective electrode length is 105 μm. Measurements are performed in a He flow cryostat, operated without the usual He heat exchange gas to enable measurements in vacuum (~1 mbar). Currents were measured with a Keithley 617 electrometer. Special care was taken regarding the wiring of the set-up, which has a leakage resistance of 1 PΩ. Prior to measuring, the samples are degassed for 48 hours at room temperature in the evacuated sample chamber. With a properly degassed sample, hysteresis of the current-voltage curves is negligible.

Figure 4 displays a set of current-voltage curves taken at different sample temperatures. The conductivity of PCPS appears to be low: At a bias voltage of 1 V, currents of less than 1 pA are measured. A room-temperature conductivity of about 10^{-10} S/cm is found. The *I-V* curves are very nonlinear. At low bias one can distinguish a linear *I-V* relation, consistent with the expected ohmic contact at the Pt-PCPS interface. At high bias the *I-V* curves display a power-law dependence, i.e., $I \propto V^a$, with a power a ranging from 4 to 10.

The data appear to be consistent with the space-charge-limited-current (SCLC) model.[8] This model is widely used to describe transport in *insulators*. The model also applies to molecular materials which generally have a large bandgap. In the absence of dopants, this leads to a low concentration of thermally excited charge carriers and a corresponding low electrical conductivity. The concentration of injected excess charge carriers now can exceed the intrinsic carrier density, a situation that is never encountered in a transport experiment on a well-conducting sample. At a certain threshold voltage V_x the concentration of injected charge equals the intrinsic charge concentration. For voltages lower than V_x the intrinsic charge carriers dominate, which leads to an ohmic current. For $V \gg V_x$, the current is space-charge limited. The amount of the injected space charge is determined by the applied voltage over the two electrodes and the capacitive coupling between the space charge and compensating charges on the electrodes. The specific form of the *I-V* curve depends on the distribution of charge traps in the bandgap. A common approach is to consider an exponential trap distribution $N(E)=N_0\exp[-(E-E_{vb})/k_BT_c]$, where N_0 is the density of trap states at the valence band edge E_{vb}, and $1/k_BT_c$ quantifies the decay rate of the trap density with energy. Charge carriers are thermally activated from the traps to the delocalized band states where they contribute to the current. An exponential trap density leads to a power-law *I-V* relation. A single analytical expression for both the ohmic and the SCLC regime can be derived by integration of Poisson's equation.[8] The result is

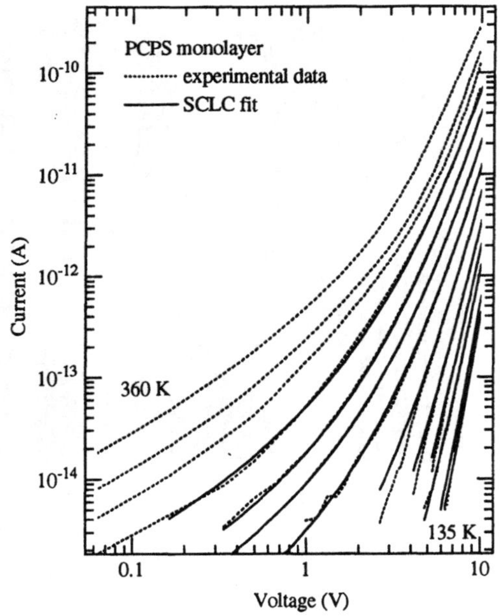

Figure 4. Current-voltage curves for a monolayer of PCPS on interdigitated electrodes. Dashed lines denote data are taken at temperatures from 135 to 360 K. Solid lines denote fits of the space-charge-limited-current model.

$$J = p\ \mu\ e\ (1+V/V_x)^{Tc/T}\ V/L \tag{1}$$

where p is the concentration of charge carriers, μ the charge carrier mobility, and L is the distance between the electrodes.

Fits of Eq. (1) - the solid lines in Fig. 4 - appear to yield excellent agreement with the *I-V* curves except at the highest temperatures. The same value of about 1500 K is found for T_c in the entire range of temperatures from 135 K to 305 K. This result, the good agreement of fits of Eq.(1) with the data, and other consistency checks,[7] justify the assumption of the exponential trap distribution. Current-voltage curves measured for samples with 9 layers of PCPS yielded similar results.

From a comparison of data taken at various spacings of the interdigitated electrodes, we find that the current is a function of the electric field V/L, in agreement with the predictions of the SCLC model for our electrode geometry. The temperature dependence of the current at fixed bias voltage appears to be not strictly thermally activated. At lower temperatures, the current is more weakly dependent on the temperature. A trap density of $\sim 10^{25}$ m^{-3} is found at room temperature, with lower values at lower temperature.

We may conclude that the space-charge-limited-currents model accounts well for the experimental results on the electrical transport in PCPS monolayers.

3. Nanofabrication of ultrasmall contacts

For transport on conjugated oligomers, nanofabrication of conducting electrodes separated by an ultrasmall (<5 nm) gap is pursued. Here we present some results obtained with a technique based on electron-beam-induced deposition of amorphous carbon. This process was pioneered in the seventies by Broers and coworkers[9] who applied it as "contamination resist".

We use this technique to grow two freestanding carbon electrodes towards each other.[10] The concept is illustrated in Fig. 5. We start with low-pressure chemical vapor deposition of a low-stress 60 nm SiN film on a Si(100) wafer. A slit is fabricated in the SiN, and Si is underetched by KOH. The slow etching rate along (111) planes results in a triangular-shape etch pit such as sketched in Fig. 5. We then sputter a thin (10-20 nm) strip of Au across the slit. In the sputtering process, some organic material (presumably diffusion-pump oil) is codeposited. We place the sample in an electron microscope with a good vacuum ($<10^{-7}$ mbar). Upon directing the 30 keV electron beam to a position near the slit in the SiN, we decompose the organic material. A deposit is formed which consists mainly of amorphous carbon.[11] Scanning the beam (cf. position 1→2→3 in Fig. 5) results in elongated free-standing whiskers with a diameter of 5 to 50 nm. A typical result is shown in Fig. 6. Note that the whisker grows perpendicular to the electron beam and parallel to the SiN membrame. From Fig. 6 it is seen that the whisker width of ~35 nm appears to be quite homogeneous when viewed from the top (i.e., along the electron beam). The side view shows some variation in the whisker height, connected to changes in the (manually controlled) scanning speed.

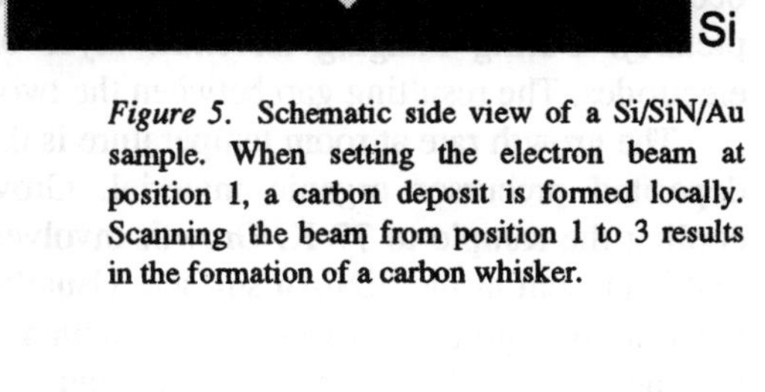

Figure 5. Schematic side view of a Si/SiN/Au sample. When setting the electron beam at position 1, a carbon deposit is formed locally. Scanning the beam from position 1 to 3 results in the formation of a carbon whisker.

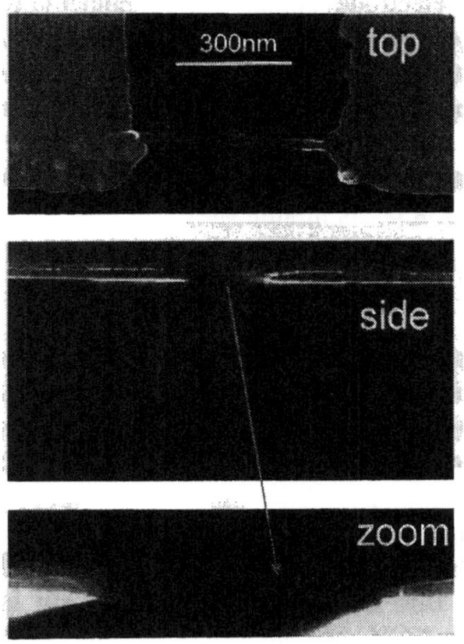

Figure 6. Top and side view of a carbon whisker grown in a slit of a SiN membrame on Si. The side view may be compared to Fig.5.

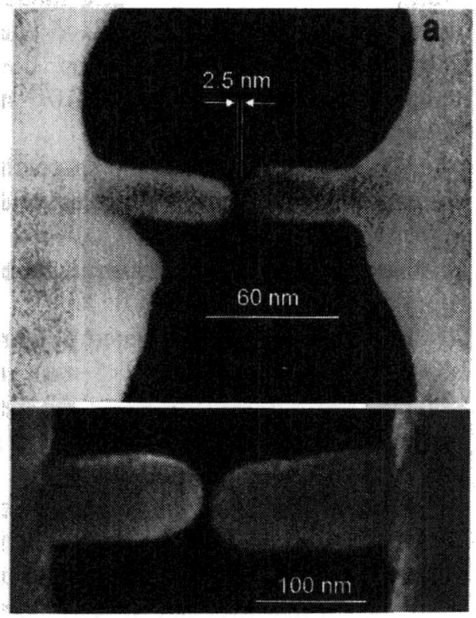

Figure 7. a. Two carbon electrodes separated by a 2.5 nm gap; b. Two electrodes coated by AuPd, separated by a 5 nm gap.

Electrodes for two-terminal transport experiments can be defined by growing two whiskers towards each other from the opposite SiN membrame edges. Carbon deposition takes place not only when the beam is focussed at a spot, but also during standard imaging. This leads to a visible increase in whisker size, and therefore to a continuous reduction of the gap size. One can stop the growth when the gap between the electrodes is a few nm. This allows the controlled fabrication of very closely spaced electrodes. Gaps down to 2.5 nm have recently been obtained. Figure 7a shows an example.

For high-resolution imaging, we may zoom in at the gap region. Figure 8 shows an image taken with a TEM. It shows two carbon electrodes on which 3 nm of Au has been sputtered. This results in Au clusters of size 5 to 10 nm (the dark spots in the picture). During imaging an extra layer of carbon (~8 nm) was grown on the electrodes. The resulting gap between the two electrodes is about 5 nm.

The growth rate at room temperature is 0.1 to 1 nm/s, dependent on the amount of deposited precursor organic material. Growth can be completely suppressed by cooling the sample to 77 K. Growth involves the decomposition of organic material that is present at the SiN/Au surface. Usually, it is *not* related to material originating from the background pressure in the vacuum system of the electron microscope. This may be concluded from the fact that - under standard conditions with a good vacuum in the electron microscope - growth does not occur if a sample has been cleaned by

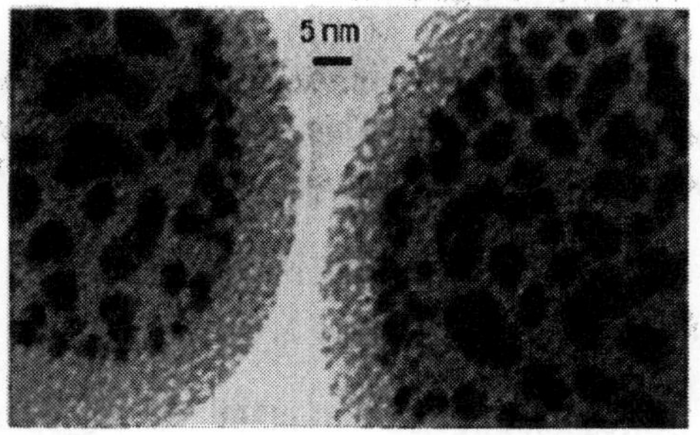

Figure 8. TEM image of the gap between two carbon electrodes which are covered with some Au clusters as well as a thin amorphous carbon layer. The gap size is about 5 nm.

aqua regia or an oxygen plasma. The temperature dependence of the growth rate is consistent with a diffusion-limited supply of material at the surface.

Carbon is an inert material that is well suited as an electrode material under various conditions. Preliminary transport measurements indicate that the resistivity is relatively high, viz., ~10^2 Ωcm.

The carbon electrodes may be coated with, for example, AuPd. This has several advantages. First, it results in a much higher electrode conductivity. Second, such electrodes may serve as a template for self assembly of the molecules, e.g., via a thiol group. AuPd is more suitable for coating than Au because it has a smaller grain size and therefore a smoother surface. AuPd is an inert material that does not form an oxide in ambient. An example of coated electrodes is shown in Fig. 7b. Here, 31 nm of AuPd has been sputtered onto two carbon electrodes that were initially separated by 16 nm. The gap has decreased to 5 nm.

The small gap sizes obtained with this fabrication scheme for contacts will allow transport experiments on molecular wires. Transport experiments on thiophene oligomers are currently underway.

4. Concluding remarks

In this contribution we have presented some steps towards electrical measurements on single molecular wires. Most important are a new route for nanofabrication of ultrasmall contacts as well as the fact that single wires of PCPS between fabricated electrodes have been obtained.

Scanning-probe microscopy can also be used for high-resolution nanofabrication. We believe that scanning-probe microscopy, deposition of organic molecules, and cryogenic electrical-transport experiments should preferrably be combined in one ultra-high-vacuum system. Experiments in this direction are underway at Delft.

From the proceedings of this workshop, it is clear that transport experiments on single molecular wires will be carried out in the coming few years. This is expected to reveal the basic physics of electronic transport in organic conductors. Additonally, it will allow to assess the feasibility of electronics at the molecular scale.

5. Acknowledgements

We acknowledge discussions with Dieter Neher, Hans Mooij, Emile van der Drift, and Lyna Sarro. This research has been funded in part by the Netherlands Organization for Fundamental Research of Matter FOM.

6. References

* Corresponding author; email: dekker@sg.tn.tudelft.nl

1. A. Aviram and M. A. Ratner, Chem. Phys. Lett. **29**, 277 (1974).
2. *Molecular Electronic Devices*, Ed. F. L. Carter (Dekker, 1986); *Molecular Electronics - Science and Technology*, Ed. A. Aviram (AIP Conf. Proc. **262**, 1992); *Introduction to Molecular Electronics*, Eds. M. C. Petty, M. R. Bryce, D. Bloor (Arnold, 1995).
3. A. Ferencz, N. R. Armstrong, and G. Wegner, Macromolecules **27**, 1517 (1994).
4. J. Wu, G. Lieser, and G. Wegner, Adv. Mater **8**, 151 (1996); J. Wu, G. Lieser, and G. Wegner, Macromolecules **29**, 1844 (1996).
5. T. J. Marks, Angew. Chem. Int. Ed. Engl. **29**, 857 (1990); B. N. Diel, T. Inabe, J. W. Lyding, K. F. Schoch, C. R. Kannewurf, and T. J. Marks, J. Am. Chem. Soc. **105**, 1551, 1983.
6. S. J. Tans, L. J. Geerligs, C. Dekker, J. Wu, and G. Wegner, preprint.
7. S. J. Tans, L. J. Geerligs, C. Dekker, J. Wu, D. Neher, and G. Wegner, unpublished.
8. M. A. Lampert and P. Mark, *Current Injection in Solids* (Acad. Press, 1970).
9. A. N. Broers, W. W. Molzen, J. J. Cuomo, and N. D. Wittels, Appl. Phys. Lett. **29**, 596 (1976); A. N. Broers, IBM J. Res. Devel. **32**, 502 (1988).
10. A. Bezryadin and C. Dekker, unpublished.
11. Y. Akama, E. Nishimura, A. Sakai, M. Murakami, J. Vac. Sci. Techn. **A8**, 429 (1990).

CO-PLANAR NANOJUNCTION TO CONNECT A SINGLE MOLECULE

V. ROUSSET [1], C. JOACHIM[1], B. ROUSSET [2], and T. ONDARÇUHU[1]

[1] *CEMES/CNRS*
29 rue Jeanne Marvig, B.P. 4347
31055 Toulouse Cedex
France
[2] LAAS/CNRS
7 Avenue du Colonel Roche
31077 Toulouse Cedex
France

Abstract. To measure the electrical resistance of a single molecule, one way is to connect it between two coplanar metallic electrodes. Our choice is to bury the metallic electrodes in the insulating SiO_2 layer, on top of a silicon substrate. Nanojunctions lower than 10 nm have been fabricated by using electron beam lithography and AuPd lift-off. SiO_2 at the same height as the metal enables us to explore the junction surface between the electrodes using an AFM. Another advantage of such device is the possibility to use the bulk silicon under the surface to "short-circuit" the nanojunction at room temperature. There are many different transport regimes through a Metal-Silicon-Metal (M-Si-M) junction depending on its size (thermionic, avalanche,..). This was used to establish that the current is localized through the nanojunction at low temperature. This makes the M-Si-M device modulable with temperature ready for the deposition of a single molecule.

1. Introduction

The objective being to measure the electrical resistance of one molecule, it must be connected between two metallic electrodes. These electrodes can be fabricated either mechanically with an STM, a breakjunction or by lithography technique. Up to now only molecules contacted by a STM have been characterized experimentally [1][2]. Compared with mechanical

C. Joachim and S. Roth (eds.), Atomic and Molecular Wires, 139-148.

techniques, lithography presents some advantages. This allows measurements on long molecules. It is difficult with a STM due to the fact that a molecule longer than 3 nm cannot stay alone perpendicular to a surface. With a breakjunction, the problem is in the detection of a molecule localized on the nanojunction. The thickness of the metallic wire prevents the AFM tip to scan it whereas for a coplanar nanojunction a grafted molecule may be imaged.
The main problem with a lithography technique is its resolution. But some experiments had proved that the electron beam lithography can allow to obtain sizes of devices in order of nm [3]. So we chose to fabricate by electron beam lithography process two coplanar metallic electrodes separated by an insulator (4). These nanojunctions must fulfill three criteria: the maximum length of a synthetized molecule is expected around 10 nm, so the junction size must be reduced to the size of one of these molecules. In order to detect with an AFM the presence of one molecule grafted on the nanojunction, the metallic electrodes and the insulator must be at the same height. By this way, an AFM tip will not be disturbed by the edges of the metallic electrodes. To agree with this criterium we have made the choice to bury the metal in the insulator. The last criterium is the expected value for the resistance of a molecule which is between 100 MΩ and 1 GΩ (5). So the device must have an electrical resistance much higher to be able to detect the current only through the molecule.

Therefore we have optimized a fabrication process based on an electron beam lithography which allow to obtain nanojunctions with a gap lower than 10 nm. It is important to notice that with this connecting technique the size and the shape of the nanojunction can be evaluated by scanning with an AFM during the electrical characterization cycle, and the presence of one molecule grafted on it will be checked by AFM and by electrical measurement.

2. Fabrication of nanojunctions with a gap lower than 10 nm

The first series of junctions were fabricated on 100 nm SiO_2 grown thermally on a silicon substrate. The pattern exposed by the electron beam consists of two electrodes 20 μm long and 1 μm wide. These two electrodes are joined by a 100 nm wide and 1 μm long wire. The junction is fabricated in the middle of the wire by stopping the electron beam over a distance measured in number of pixels jumps. To obtain junction with electrodes and insulator at the same height, the AuPd metal is buried on the SiO_2 by using a wet etching. With this technique, junctions with a gap of 40 nm have been obtained with a yield of 40%. By modulating the number of pixel jumps, nanojunctions with a gap of 8 nm have been fabricated with still SiO_2 between the two metallic electrodes *(Figure 1)* [4]. But compared with the 40 nm junctions, the yield is smaller, around 5%. This may be attributed to the etching step: the etching rate is high and in spite of the dilution, the etching solution may destabilize the resist protecting the SiO_2 in the gap. At the end of the fabrication process, this results in a continuous metallic wire.

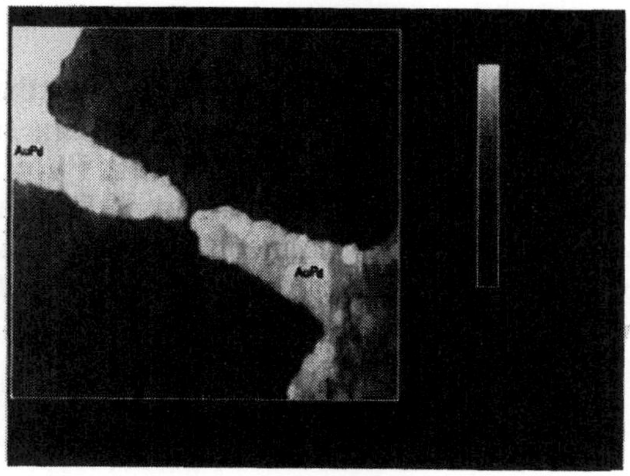

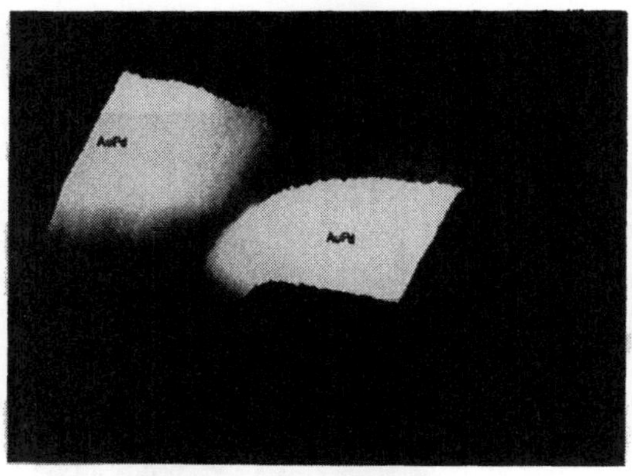

Figure 1. **a)** Top view AFM image of a nanojunction with a 8 nm gap fabricated on 100 nm of SiO_2 grown thermally on a silicon substrate. The two extremities of the 20 μm long and 1 μm wide AuPd electrodes have been scanned. The nanojunction is present in the middle of the 1 μm long and 120 nm wide AuPd wire. **b)** 3D AFM image view with a smaller scan of this nanojunction. The extremities of the AuPd wire delimit the nanojunction. There is still SiO_2 between them. The metal is higher than 3 nm compared to the SiO_2.

During the first electrical characterization, these devices were very often destroyed by electrostatic discharges. Therefore, for the second series of nanojunctions, we have eliminated the SiO_2 layer between the metallic electrodes and the silicon substrate. By this way, the properties of the doped silicon are used at room temperature to prevent the destruction of the device. At smaller temperature, the silicon bulk conductivity is suppressed so the current's lines can be localized through the nanojunction. This modification led us to optimize the fabrication process on this new substrate in order to obtain nanojunctions with a gap lower than 10 nm again.

The device are now fabricated on 10 nm SiO_2 grown thermally on a P-doped silicon substrate. The pattern is exposed with the same dose, only the development step have been investigated again. By controlling the 10 nm wet SiO_2 etching, nanojunctions with a gap of 8 nm in direct contact with the P-doped silicon have been obtained *(Figure 2).* Like for junctions fabricated on SiO_2 substrate, the fabrication yield for junction lower than 10 nm is small : 4 successful nanojunctions per 22 scanned by an AFM.

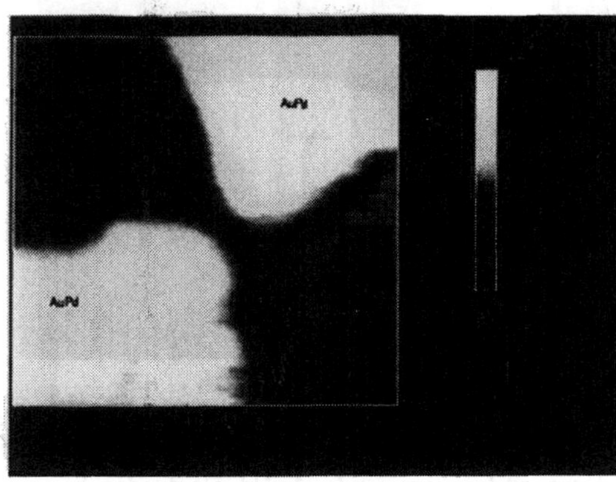

Figure 2. Top view AFM image of a 8 nm nanojunction obtained with the new fabrication process on 10 nm of SiO_2. The extremities of the AuPd wire are clearly delimited.

Before grafting a molecule on the nanojunction, the current-voltage characteristics of the M-Si-M nanodevice must be studied at different temperatures to check that at low temperature the current go really through the nanojunction and with its intensity smaller enough to allow the electrical measurement of one molecule.

3. Electrical characterization of nanojunctions in direct contact with silicon substrate

To be able to measure the current through a nanojunction, it must be connected to the apparatus. So with a classical photolithography process, the nanojunction is contacted by bigger metallic pads. The first deposited layer is 50 nm of Ni to make a good contact with the silicon substrate, and 600 nm of gold have been deposited over by sputtering. The complete device consists of three M-Si-M junctions in parallel with differents sizes *(Figure 3)*. The first one Ni/Au-Si-Ni/Au has a length of 20 μm, an other one separates the two AuPd electrodes and is 1 μm long. The last one has the size of the nanojunction.

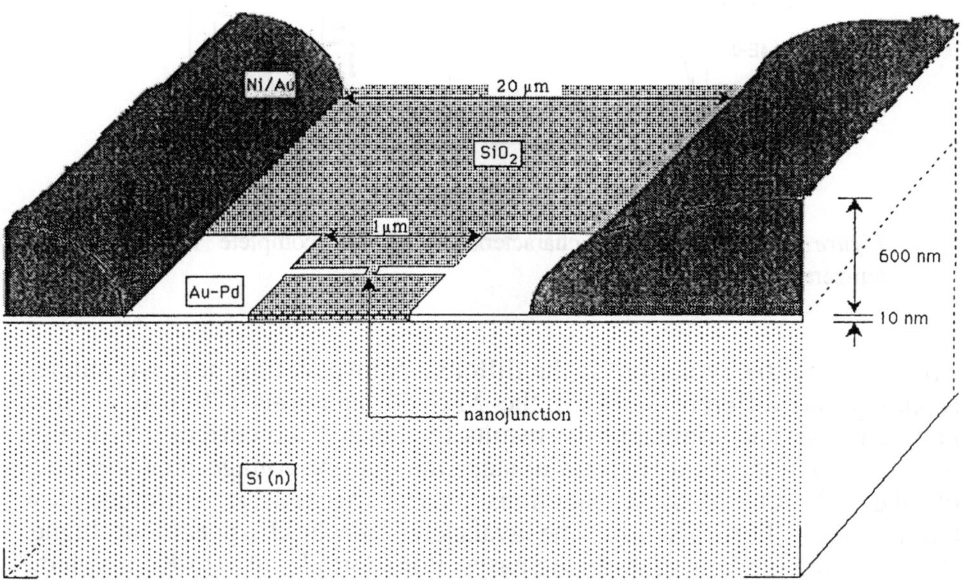

Figure 3. Picture of the complete M-Si-M nanodevice. The two Ni/Au pads contacting the AuPd structure fabricated with an e beam lithography are separated by 20 μm and are formed with a classical photolithography process. Both Ni/Au and AuPd are in direct contact with the silicon substrate. Therefore, there are differents sizes of M-Si-M junctions : Ni/Au-Si-Ni/Au, AuPd-Si-AuPd, and the last one with the size of the nanojunction.

Figure 4 shows current-voltage characteristics for the complete device at room temperature. There are clearly different transport regimes. For low voltage, up to 0.2 volt, the current increases rapidly with the voltage. It's due at the thermionic emission of the electrons through the schottky barrier. The schottky barrier height measured on this device with a classical technique gives 0.42 eV for the effective barrier [6]. As the voltage increases, higher than 4 volt, the current increases rapidly too. But in this case, it is due to the multiplication of carriers, the avalanche regime is reached.

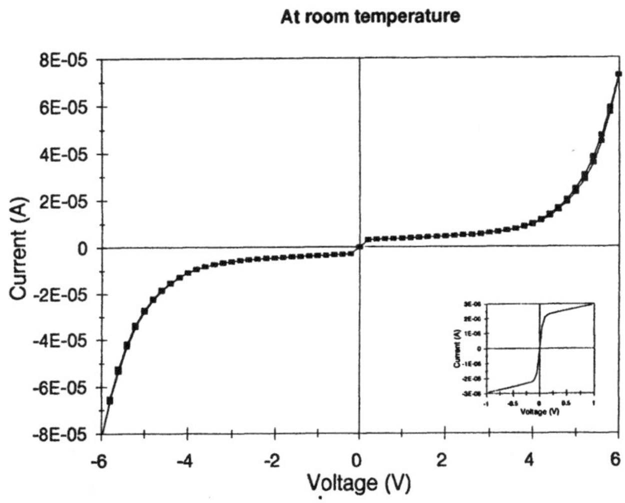

Figure 4. Current-voltage characteristics of the complete device at room temperature.

To determine the evolution of the different transport regimes, we studied separatly the different sizes of the M-Si-M junctions (20 μm, 1 μm, nanojunction) and compared the results to check if the current go through the nanojunction. As a first trivial step, the current-voltage characteristics of the 20 μm M-Si-M junctions were checked at different low temperatures. For this size, the avalanche threshold appears at higher voltage when the temperature decreases. For example, at 100 K, a voltage of -3 volt must be applied to obtain 2 nA which is reached at -1 volt at 200 K. Moreover at low temperature, for a large range of voltage, the current is lower than 1 nA. As it is well know, transport in this 20 μm junctions is suppressed for small voltage by decreasing the temperature. As a second step, we have tested 1 μm AuPd-Si-AuPd junctions. We have observed the same evolution in the I(V) curves during the temperature decrease compared to the 20 μm. The main difference is in the value of the voltage threshold which appears at smaller value for the 1 μm junction M-Si-M. At low voltage and low temperature, the current through this size of junction can be suppressed too. Therefore, we are now able to characterize a nanojunction contacted by these big metallic pads at low temperature when all the bigger M-Si-M junctions are "short-circuited" by the nanojunction.

In the last step of electrical characterization cycle, devices with nanojunctions of 80 nm and 30 nm have been tested. In *Figure 5*, we compare I(V) curves at 200 K obtained for two sizes of M-Si-M junctions: 20 μm and the complete device with a 80 nm nanojunction. At this temperature, for applied voltage lower than about 1.5 volt the current through 20 μm M-Si-M junction is lower than pA. The I(V) measurement on the complete device is therefore characteristic of the 80 nm nanojunction presence. At -1 volt the current is approximately one order of magnitude higher than for the 20 μm junction. For this size of M-Si-M junction, we have noted that the breakdown threshold has a negative temperature coefficient, as the temperature decreases the breakdown threshold increases. This shows a different transport regime compared to the biggest M-Si-M junctions with a positive temperature coefficient like in a classical avalanche regime.

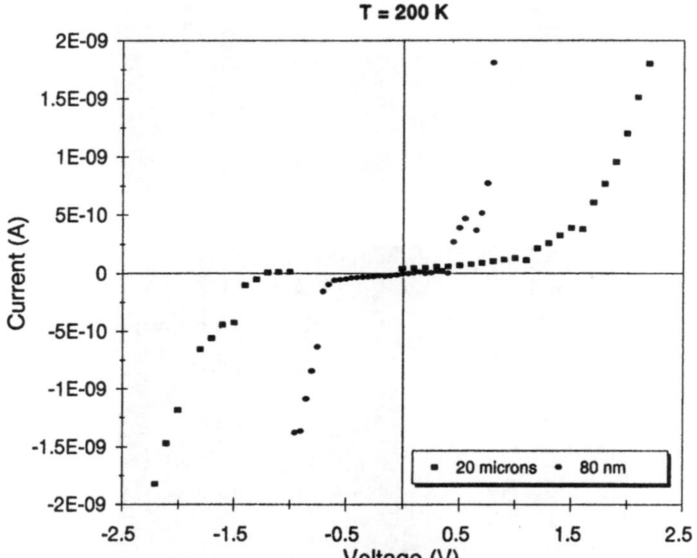

Figure 5. Current-voltage characteristics at 200 K measured on two different devices: one with only the 20 μm M-Si-M junctions and the other on all the device with a 80 nm nanojunction.

Therefore, different transport regimes depending on junction sizes have been established at low temperature and we have checked that the current go through the nanojunction. The 80 nm nanojunction has been imaged before and after electrical measurement to be sure that it is still present. The 3D AFM images *(Figure 6)* shows no change in shape and size of the nanojunction. We can note that the roughness of the silicon dioxyde between the metallic electrodes may allow the detection with an AFM of one molecule localized on the nanojunction.

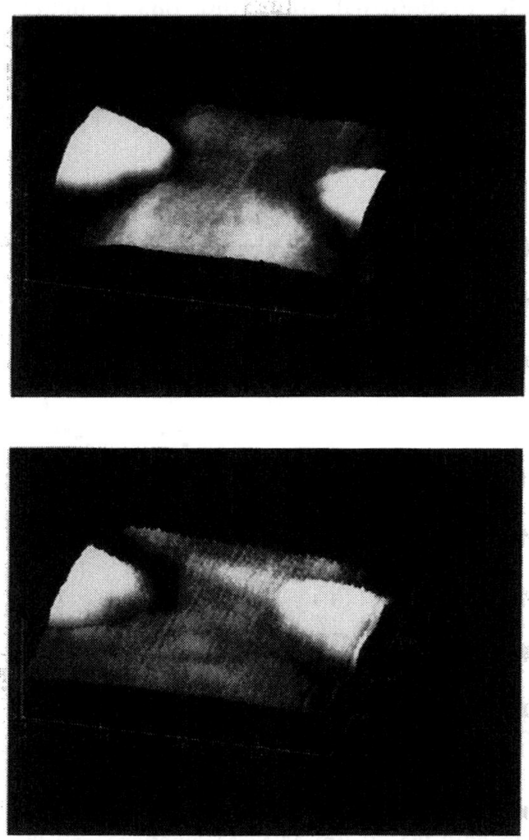

Figure 6. 3D AFM image view of the 80 nm nanojunction before a) and after b) the current-voltage measurements.

For small voltage to determine at which temperature the current through a nanojunction can be minimized, we have tested a 30 nm nanojunction at different temperatures. At 100 K *(Figure 7)* there is a large range of voltage, around 1 volt, where the current is smaller than 1 pA which gives a resistance larger than the Tera Ohm.

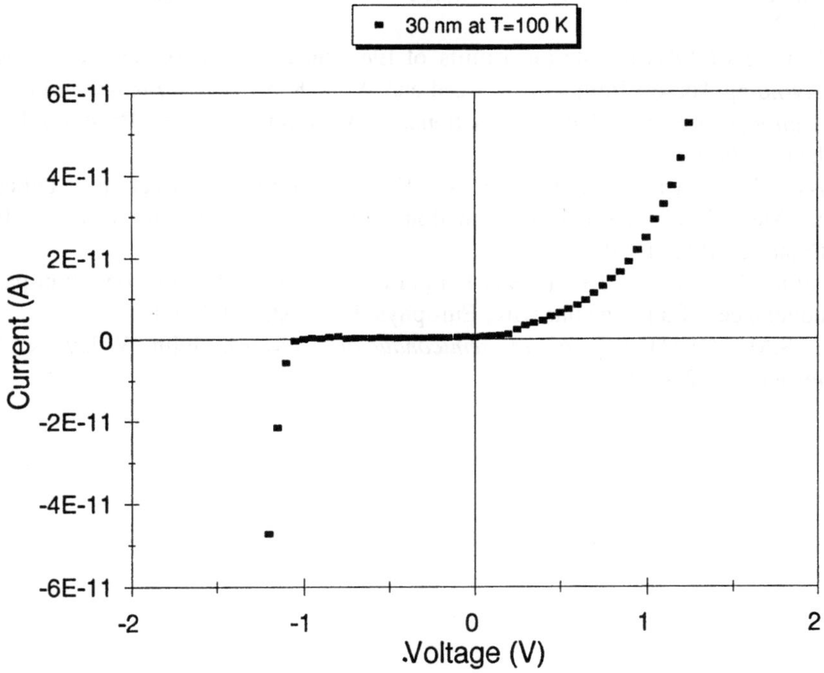

Figure 7. Current-voltage characteristics of a 30 nm nanojunction at 100 K.

4. Conclusion

We fabricate co-planar nanojunctions with a gap lower than 10 nm. Size and shape of the nanojunction are determined by an AFM during the electrical measurement cycle to fulfill that there is no destruction. The small difference in height (3 nm) between the metallic electrodes and the SiO_2 allows to scan the junction and will allow the detection of one molecule grafted on the nanojunction. Moreover, the device is characterized, at all temperature showing different transport regimes. The important point is that at low temperature the current's lines are localized through the nanojunction. The value of the nanojunction resistance obtained at small temperature (Tera Ohm at 100 K) is much higher than the value expected for one molecule [5].

5. References

1. Joachim, C., Gimzevski, J.K., Schlittler, R. R., Chavy, C. (1995) Electronic Transparence of a Single C_{60} Molecule, Phys. Rew. Lett. **74**, 11, 2102-2105.
2. Andres, R.P., Bein, T., Dorogi, M., Feng, S., Henderson J.I., Kubiak, C.P., Mahoney, W., Osifchin, R.G., Reifenberger, R. (1996) "Coulomb Staircase" at Room Temperature in a Self-Assembled Molecular Nanostructure, Science **272**, 1323-1325.
3. A.N. Broers (1989) Resolution Limits of Electron Beam Lithography and Methods for Avoiding These Limits, M.A. Reed and W.P. Kirk, *Nanostructure physics and fabrication*, Proceeding of the international symposium, Academic Press Publishers, College Station Texas, pp 421-430.
4. Rousset, V., Joachim, C., Rousset, B., Fabre, N. (1995) Fabrication of buried co-planar Metal-Insulator-Metal nanojunctions with a gap lower than 10 nm, J. Phys. III France **5**, 1983-1989.
5. Joachim, C., Vinuesa, J.F. (1996) Length dependance of the electronic transparence (conductance) of a molecular wire, Europhys. Lett., **33**, 8, 635-640.
6. Sze, S.M. (1981) *Physics of semiconductors devices*, John Wiley & Sons Publishers, pp 284-286.

CONDUCTANCE QUANTIZATION IN METALLIC NANOCONTACTS: EXPERIMENTAL RESULTS AND NOVEL THEORETICAL ASPECTS

N. GARCÍA, J.L. COSTA-KRÄMER, A. GIL,
P. GARCÍA-MOCHALES, AND P.A. SERENA

Física de Sistemas Pequeños (CSIC-UAM)
Universidad Autónoma de Madrid, C-III
Cantoblanco, E-28049-Madrid, Spain

ABSTRACT. In this paper we present results showing the main features of the conductance quantization in metallic contacts under uncontrolled and controlled conditions. We present histograms formed by thousands of single STM measurements, revealing the universality and reproducibility of this phenomenon. The analysis of such histograms reveals some interesting features, explained by tight-binding calculation in terms of the presence of disorder within the nanowire. We have determined the possible chaotic distribution of the electronic states within the nanocontact by studying the distribution of elapsed times in a given conductance level. We also present new results showing the existence of light emission from the nanocontact.

1. Introduction

During the last decade, there has been an increasing interest in the fabrication and characterization of nanoscale conducting systems due to their wide range of applications in communication and computing technologies [1]. In 1987 it was reported that the electron conductance through small constrictions created in a two-dimensional electron gas is quantized in units of the quantum of conductance $G_0 = 2e^2/h$ [2], where e is the electron charge and h is the Planck constant. The advent of the scanning tunnelling microscopy (STM) [3] has opened the possibility of studying with extreme accuracy the evolution from contact to tunnel regimes when two electrodes (tip and sample) separates each other. Also in 1987, it was shown [4] that when the tip approaches towards the sample surface an abrupt jump from tunnel to contact occurs being this jump associated to a resistance of ≈ 10 kΩ.

With the continuous improvement of experimental techniques the phe-

C. Joachim and S. Roth (eds.), Atomic and Molecular Wires, 149-158.

nomenom of the quantization of the conductance in metallic contacts has been studied from two different prospects: STM based methods [5-8] and the mechanically controllable break junction (MCBJ) method [9,10]. Both methods have proven that at low and high temperatures the conductance quantization appears for several metallic species, although they involve sophisticated equipment and control software. In a very recent experiment [11], we found the same conductance quantization features when placing two macroscopic wires in contact and making them vibrate in order to form and break of contacts. With this simple setup it has been possible to analyse the conductance quantization for a wide variety of metallic macroscopic contacts under different environments [12].

In the present work, we describe the main features of the quantization of the conductance in both limits, controlled and uncontrolled experiments, and present new results where the statistical analysis reveals the universality of this phenomenon. The obtained histograms were analysed and some interesting features have been explained in terms of the presence of disorder within the nanowire by means of tight-binding calculations [13]. The statistical analysis will allow to connect the duration of the conductance steps with the energy separation between electronic states formed within the nanowire during the nanocontact formation or breaking. From these measurements we have determined the chaotic distribution of the electronic states within the nanocontact [12]. The last section will show experimental results indicating that there is light emission from the nanocontact during its breaking process [14].

2. From table-top to controlled experiments

In a recent experiment [11] we have demonstrated that the quantization of the conductance in metallic contacts can be obtained with a simple setup: two metallic wires, a power supply and an oscilloscope. By tapping on top of the table we are able to form and break contacts between both wires due to the induced vibrations. In Figure 1 we show a typical conductance curve obtained with this method. This kind of curves, showing the signature of the contact breaking process, are indistinguishable from those obtained with STM and MCBJ techniques [5-10]. This experiment clearly shows that, during the last stages of the breaking process of a macroscopic contact, the formation of a nanowire is taking place. With the same method we have found that the quantization also occurs in a wide variety of environmental conditions (air, UHV, liquids) as well as in liquid metallic species [12]. However, these experiments do not provide a suitable way of determining the physical properties of the metallic nanocontacts. We have performed similar measurements but taking control on the approaching between both macro-

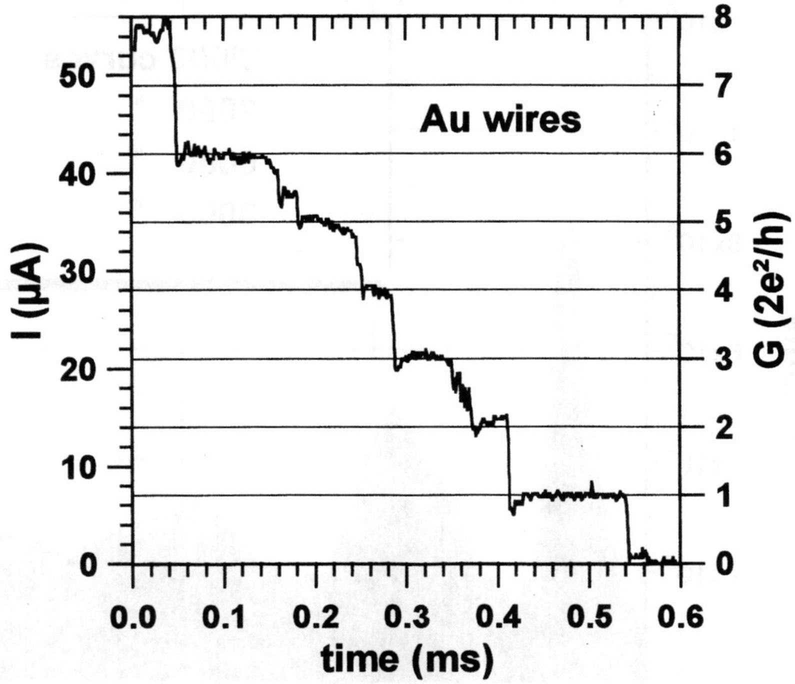

Figure 1. Single conductance measurement obtained with a table-top setup, showing the evolution of the intensity and the conductance at the last stages of the separation between two gold macroscopic wires. Applied voltage between both wires is 90.3 mV.

scopic wires using controlled piezodisplacement techniques in UHV [12].

In order to perform a more rigorous study of the conductance quantization, we have carried out several experiments using a STM-like setup, thus being able of controlling the speed of the separation between both macroscopic Au electrodes [15]. The experiments were performed at RT in air. The current flowing through the contact is measured with an I-V converter connected to a digital oscilloscope. The oscilloscope allows the construction and display in real time of conductance histograms, where all the experimental curves are used in the statistical analysis. In Figure 2 we show four different histograms corresponding to 3000, 6000, 9000, and 12000 different samples (single breaking contact experiments). In this particular case, we have substracted a residual resistance of 490 Ω from the original data. This resistance shifts the first peak to its expected value. The origin of this resistance is explained in the next Section. This histogram shows that there is no need of neglecting experimental curves on the basis of some *a priori* criterion to get a clear signature of the conductance quantization. Similar

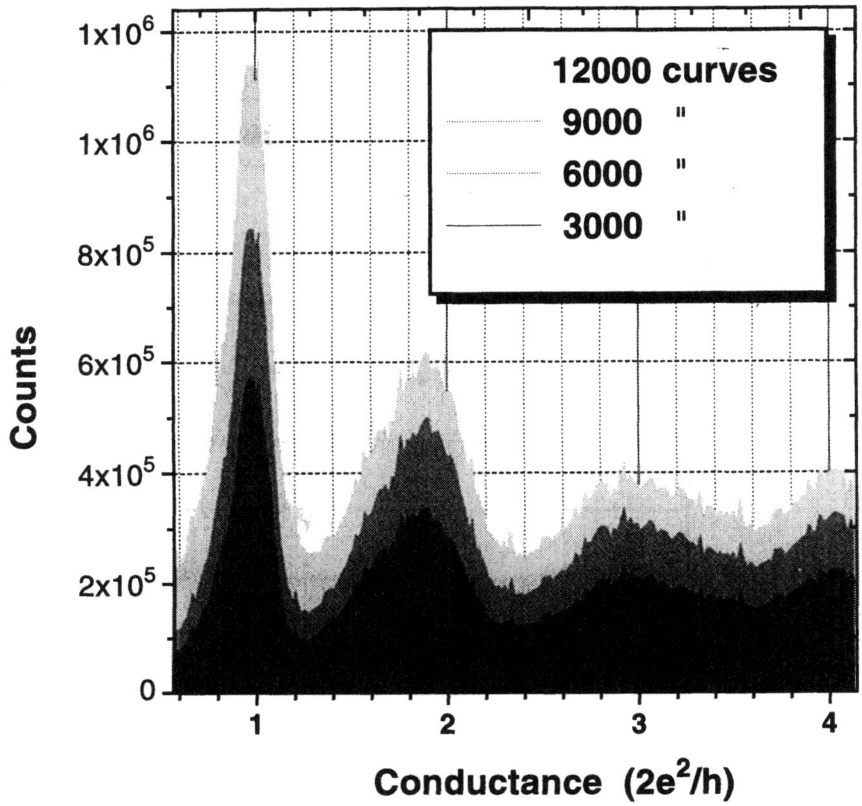

Figure 2. Conductance histograms for gold nanocontacts obtained at room temperature and in air. The applied voltage between both electrodes is 90.3 mV. The electrodes were separated at 89000 Å using a STM system. Histograms accumulate data over N= 3000, 6000, 9000 and 12000 single measurements.

results were obtained for copper and silver contacts, where histograms were composed of more than 10^5 curves [15]

3. The effect of the disorder within the nanowire

In order to explain the origin of the residual resistance needed to shift the first conductance plateau to its theoretical value G_0, we have assumed that the presence of disorder within the nanowire is a reasonable possibility. To check this hypothesis, we have performed tight-binding calculations [13] where disorder is included following Anderson model [16]. The nanocontact is simulated by a 2D wire with a region where disorder is included by changing the atomic orbital eigenenergies at random between $-D/2$ and

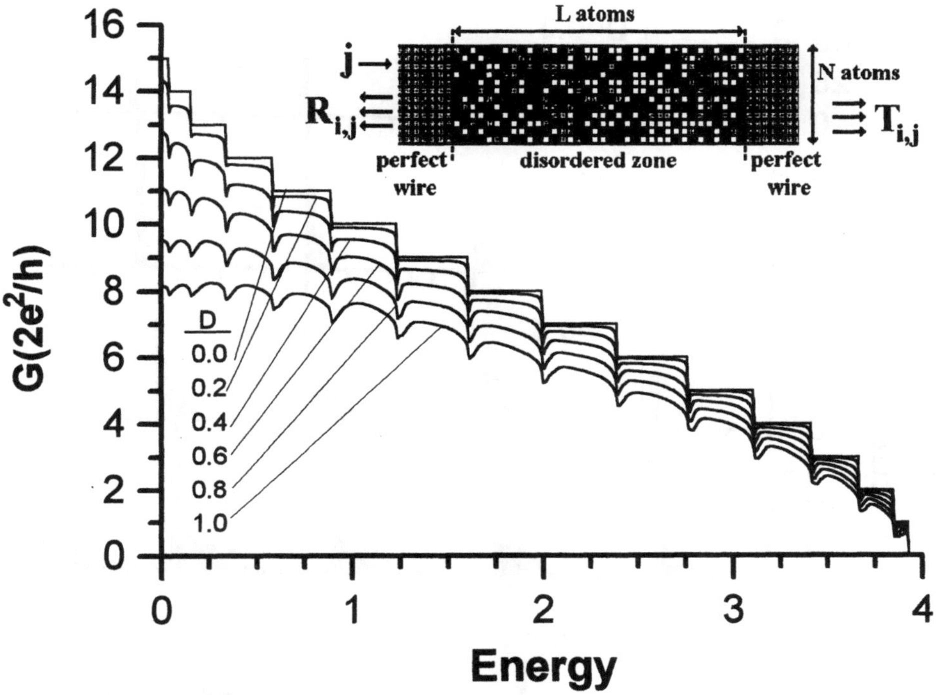

Figure 3. Calculated conductance G as a function of the Fermi energy E of the incoming electrons, for a wire of length L=15 and width W=15, and for different values of the disorder parameter D= 0, 0.2, 0.4, 0.6, 0.8, and 1.0. The conductance was obtained after averaging over 250 different curves for each disorder parameter. Upper part: scheme of the calculational model.

$D/2$ (D is the disorder parameter). From our calculations we obtain the transmission probabilities $T_{i,j}$, the probability that an incident electron in the left channel j emerges in the right one i. With these quantities we evaluate the wire conductance using the two-probe Landauer formula $G = \frac{2e^2}{h} \sum_{i,j} T_{i,j}$ [17].

In Figure 3 we show results for a disordered wire of length $L = 15$ and width $N = 15$. We have computed the average conductance over an ensemble of 250 realizations for each disorder parameter D. The x axis represents the Fermi energy of the system, and its variation permits to change the population of the incident modes. We can notice that the conductance curve degenerates for increasing disorder. Furthermore, well defined dips appear between two adjacent conductance plateaus. These dips are similar to those experimentally observed in many conductance measurements (see, for instance marked dips in Figure 1). With those curves shown in Figure 3 we can compose similar histograms to those of Figure 2. In Figure 4 we

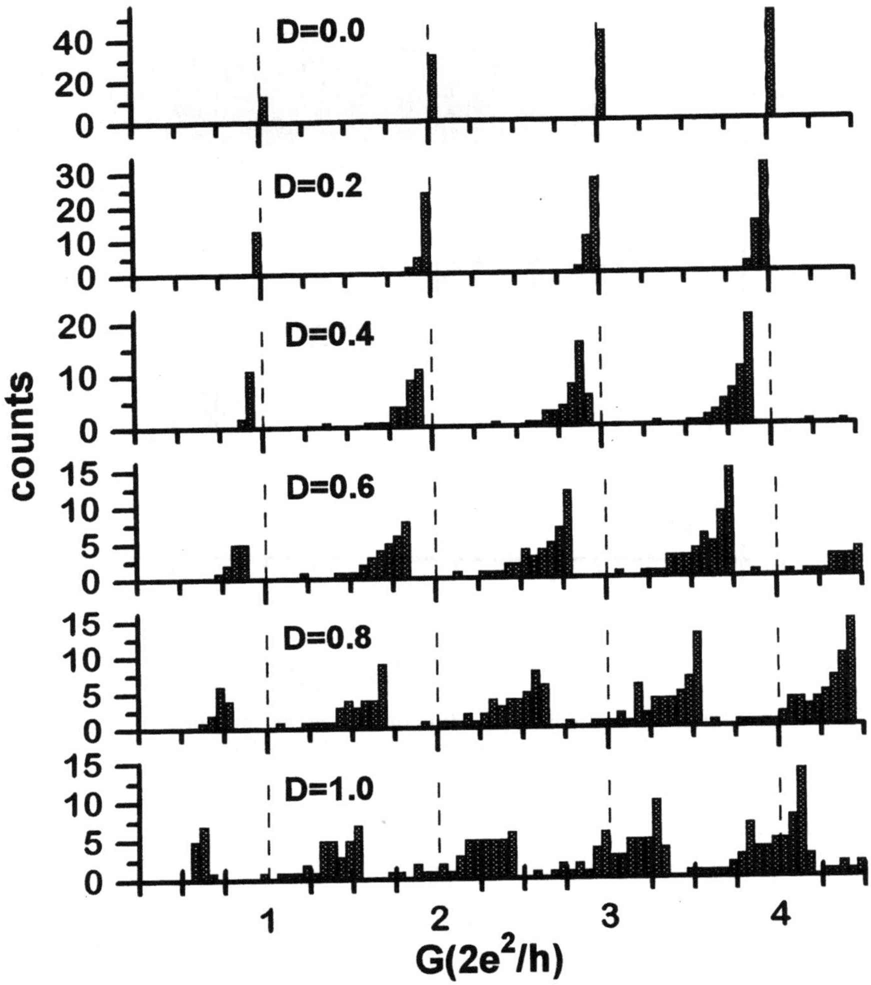

Figure 4. Conductance histograms obtained from the curves shown in Figure 3.

present these calculated histograms for different disorder parameters. Notice that for perfectly ordered wires, $D = 0$, we have a histogram formed by δ-functions, whereas for increasing disorder the conductance peaks shift to lower values. We can see that the peak shift is proportional to the conductance channel, and this can be interpreted in terms of a constant resistance. Therefore a small amount of disorder explains the presence of dips in the conductance curves as well as the origin of the residual resistance.

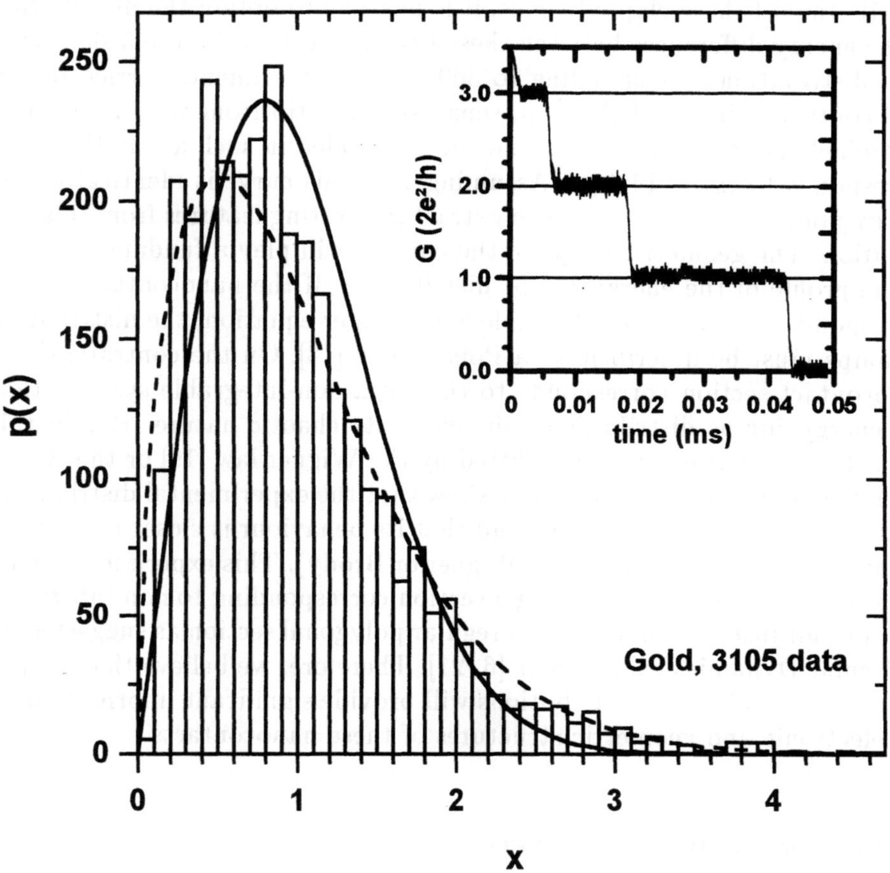

Figure 5. Histograms showing the distribution of the duration of the first conductance plateau for macroscopic contacts formed between two gold wires. The x-axis represents the time (x=1 corresponds to 14.4 μs). Solid and dashed lines correspond to theoretical distributions of states (Wigner and Brody, respectively) for chaotic quantum systems. The inset shows a single conductance experiment carried out under the conditions described in Figure 2.

4. Chaotic distribution of nanowires quantum states

When analysing an experimental curve as that shown in Figure 1, a very interesting quantity is related with the duration of the different quantized conductance values. By simple arguments, it is possible to associate the elapsed time of the conductance in a given quantum state with the separation between two subsequent quantum levels [12]. In particular, the time elapsed by the conductance curve in the value G_0 is proportional to the energy difference between the first and second quantized levels inside the nanocontact at the latest stages of the breaking process. By studying the distribution of these elapsed times, it is possible to obtain the distribution of the energy differences between these two eigenvalues. Figure 5 shows the time distribution corresponding to 3000 single experiments carried out in gold contacts with a STM setup, separating the tip from the sample at a controlled speed of 89000 Å/s. The curve is scaled in such a way that x=1 corresponds to $\Delta t = 14.4\ \mu s$. Assuming that this curve is identical to the energy jumps distribution, we can extract interesting features from this distribution. The geometric shape of the contact will play a fundamental role in the profile of the energy jumps distribution. If the nanocontact section corresponds to that of an integrable Schrödinger equation, the distribution of jumps must be described by a Poisson law [18]. On the contrary, if the nanocontact section corresponds to that of a non-integrable system, thus the energy jumps distribution will reflect its chaotic nature [19]. In this case, the distribution should be fitted by the Wigner law [19] or the Brody formula [20]. In Figure 5 we clearly show that the experimental distribution does not follow the Poisson law, and that its behaviour is closer to that described by chaotic distributions (Wigner or Brody). This experiment points out towards nanocontacts having a section corresponding to non-integrable systems, for instance that of a non regular polygonal section as suggested by Molecular Dynamics calculations [6, 21]. Therefore, we believe that deeper analysis of this kind of distributions will provide significant information on the electronic and geometric structures of these nanocontacts.

5. Light emission from nanocontacts

This last Section addresses a different aspect of the physical phenomena taking place in a nanocontact. We have performed several experiments showing that it is possible to obtain light emission from a nanocontact when relatively high voltages are applied [14]. In Figure 6 we show the strong correlation between a gold nanocontact breaking process and the presence of photons in the photon counter. In this particular case an applied voltage of 2.6 V was used, although this phenomenom appears above ≈1.0 V. We have demonstrated that this light emission happens for different metallic species (Au, Ag, Pt, W, Al) and under different environments (air, oxygen and helium atmospheres). These nanowires are as little bulbs that are switched on during the breaking process of the nanowire. The explanation for this light emission points out towards a mechanism in which a very hot electron gas is in a non-equilibrium situation with the phonons of the nanowire, due to the small electron-phonon coupling in such systems, and therefore there is blackbody radiation due to the presence of hot electrons [14].

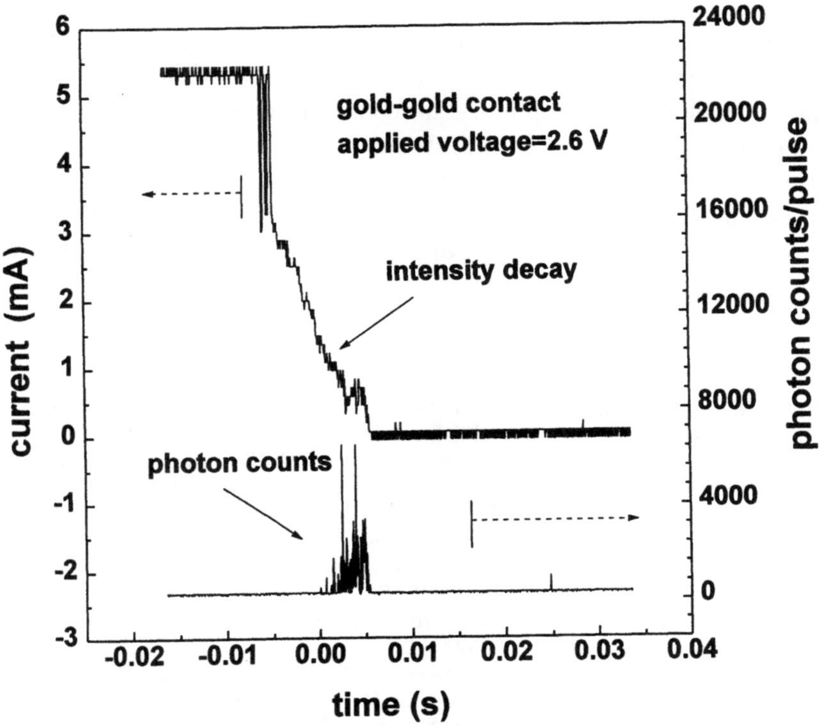

Figure 6. Current intensity and simultaneous photon emission measurements during the breaking process of a gold contact formed between an Au tip and an Au plate for an applied bias voltage of 2.6 V. These measurements were carried out in air at room temperature.

6. Concluding remarks

In this work we have presented several experimental and theoretical results illustrating the quantization of the conductance in nanocontacts. We have shown experimental current decay curves based on table top experiments as well as results where there is control on the breaking process. Histograms formed by a huge number of single experimental data permit us to conclude that this phenomenom is reproducible, giving rise to well defined histograms characterizing the contact species. The analysis of the experimental curves has been completed with tight-binding calculations, in order to show that a small amount of disorder within the nanocontact will explain some phenomena as the presence of dips in the conductance curves or the shift of the conductance histogram peaks, usually connected to a residual resistance. By means of a detailed analysis of the distribution of times corresponding to different quantized conductance values, we have been able of study the kind of geometric sections for the nanocontact at the latest stages of the breaking process, indicating the connection of this analysis with quantum chaos theories. Other experimental result we have presented concerns with the light emission from nanocontacts when voltages higher than 1 V are applied. These experiments demonstrate the rich phenomenology existing in these systems.

Authors would thank helpful discussions with A. Correia and M. Marqués. This work has been supported by the Spanish DGICyT PB94-0151 and CICyT MAT94-1456-CE projects, and the EU Programmes HCM and ESPRIT.

References

1. Welland, M.E. and Gimzewski, J.K. (eds.) (1995) *Ultimate Limits of fabrication and Measurement*, Kluwer Academic Publishers, Dordrecht.
2. van Wees, B.J., van Houten, H., Beenakker, C.W.J., Williamson, J.G., Kouwenhoven, L.P., van der Marel, D., and Foxon, C.T. (1988) *Phys. Rev. Lett.* **60**, 848; Wharham, D.A., Thornton, T.J., Newbury, R., Pepper, M., Ahmed, H., Frost, J.E.F., Hasko, D.G., Peacock, D.C., Ritchie, D.A., and Jones, G.A.C. (1988) *J. Phys. C* **21**, L209.
3. Binnig, G., Rohrer, H., Gerber, Ch., and Weibel, E. (1982) *Phys. Rev. Lett.* **49**, 57.
4. Gimzewski, J.K. and Möller, R.M. (1987) *Phys. Rev. B* **36**, 1284.
5. Pascual, J.I., Méndez, J., Gómez-Herrero, J., Baró, A.M., García, N., and Vu Thien Binh (1993) *Phys. Rev. Lett.* **71**, 1852.
6. Pascual, J.I., Méndez, J., Gómez-Herrero, J., Baró, A.M., García, N., Landman, U., Luedtke, W.D., Bogacheck, E.N., Cheng, H.-P. (1995) *Science* **267**, 1793.
7. Agraït, N., Rodrigo, J.G., and Vieira, S. (1993) *Phys. Rev. B* **47**, 12345.
8. Olesen, L., Laegsgaard, E., Stensgaard, I., Besenbacher, F., Schiotz, J., Stoltze, P., Jacobsen, K.W., and Norskov, J.K., (1994) *Phys. Rev. Lett.* **72**, 2251.
9. Muller, C.J., van Ruitenbeek, J.M., and de Jongh, L.J. (1992) *Phys. Rev. Lett.* **69**, 140.
10. Krans, J.M., van Ruitenbeek, J.M., Fisun, V.V., Yanson, I.K., and de Jongh, L.J. (1995) *Nature (London)* **375**, 767.
11. Costa-Krämer, J.L., García, N., García-Mochales, P., and Serena, P.A. (1995) *Surf. Science* **342**, L1144.
12. García, N., Costa-Krämer, J.L., Serena, P.A., García-Mochales, P., and Bitar, L. (1996) in *Physical and Supramolecular Chemistry*, edited by Echegoyen, L., and Kaifer, A.E., Kluwer Academic Publishers, Dordrecht.
13. García-Mochales, P., Serena, P.A., García, N., and Costa-Krämer, J.L., (1996) *Phys. Rev. B* **53**, 10268.
14. Gil, A., Sharonov, M., García, N., and Calleja, J.M., to be published.
15. Costa-Krämer, J.L., and García, N., to be published.
16. Anderson, P.W. (1958) *Phys. Rev.* **109**, 1492.
17. Landauer, R. (1957) *IBM J. Res. Dev.* **1**, 233; Landauer, R. (1989) *J. Phys. Condens. Matter* **1**, 8099.
18. Berry, M.V., and Tabor, M. (1977) *Proc. Roy. Soc,. London* **356**, 375.
19. Bohigas, O., Gianoni, M.J., and Schmit, C. (1984) *Phys. Rev. Lett.* **52**, 1; McDonald, S.W., and Kaufman, A.N., (1979) *Phys. Rev. Lett.* **52**, 1189; McDonald, S.W. and Kaufman, A.N. (1979) *Phys. Rev. Lett.* **52**, 1189.
20. Brody, T.A. (1973) *Lett. Nuovo Cimiento* **7**, 482.
21. Landman, U., Luedtke, W.D., Burnham, N.A., and Colton, R.J. (1990) *Science* **248**, 454; Landman, U. and Luedtke, D. (1991) *J. Vac. Sci. Technol* **9**, 414.

ELECTRICAL RESISTANCE OF ONE OR TWO ATOMS

ALI YAZDANI, D. M. EIGLER
IBM Research Division
Almaden Research Center
650 Harry Road,
San Jose, CA 95120, USA

AND

N. D. LANG
IBM Research Division
Thomas J. Watson Research Center
Yorktown Heights, NY 10598, USA

1. Introduction

What is the nature of the electrical transport in atom-sized devices?

Recent advances such as the scanning tunneling microscope (STM) and the mechanically controllable breakable junctions (MCB) have made it possible to investigate experimentally the above question. So far much of the effort in understanding transport in this regime has been to rely on the by-now-familiar Landauer picture of quantized conduction channels in narrow constrictions. This idealized view of one-dimensional transport stems naturally from the importance of the wave nature of the electron in its transport through constrictions with dimensions comparable to the electron's wavelength. The electrical conductance of such structures are predicted to be quantized in units of $2e^2/h \sim 1/12.9$ kilohms [1]. While these concepts of electrical transport have been very successful in capturing the physics of electron motion in mesoscopic conductors such as 2D electron gas devices realized in semiconductor heterostructures [2], their applicability on the atomic scale may be questioned. One question that can be asked is: How strong a role do the electronic properties of a single atom play in determining the resistance of a nanometer-sized object?

The idea that the conduction process through a chain of just a few atoms is determined by the character of the atomic orbitals of the atoms

C. Joachim and S. Roth (eds.), Atomic and Molecular Wires, 159-168.

and their interaction with the electrodes is rather intuitive to understand. However, prior to the work described here (reported first in [3]) there are no examples of atomically controlled experiments performed together with detailed electron density calculations to demonstrate this idea. To address these question we used the STM to construct and measure the resistance of atomic wires consisting of one or two Xe atoms. Comparing the experimental results with fully self-consistent calculations of the resistance, we show that the resistance for such wires is mainly determined by the state-density contribution of the atomic resonances at the Fermi level (E_F's) of the electrodes.

2. Making Contact to a Single Atom

Recently a number of techniques such as STM and MCB have been used to make narrow metal constrictions. A number of these studies have shown that bringing two metal surface close together results in a residual jump-to-contact resistance of order $h/2e^2$ ~12 kilohms, the value expected for a 1D conductor with one ideal propagating channel of conductance. In a typical metal the electron wavelength at the Fermi energy is of the order of the interatomic spacing; hence it has been argued that the channel with 12.9 kilohm resistance must have dimensions comparable to that of a single atom. Apart from the experimental measurement, there are simulations of the mechanical interaction between two metal surfaces which support the one-atom picture for the metallic contacts [4].

Despite the uncertainty in the number of atoms in between the macroscopic electrodes, previous experiments have shown the resistance of these narrow constrictions to be material-dependent, hence suggesting the importance of the atomic character [5]. In the work described here, we have used the STM's ability to image and manipulate individual Xe atoms to construct our atomic wires. In making contact to single atoms, the STM offers the imaging capability which is an important advantage in characterizing the atoms that are in between the two electrodes; in this case, a polycrystalline W wire as the STM tip and a Ni(110) single-crystal surface.

3. Xe Atom in the STM Junction

Xe atoms adsorbed at low temperature (T~ 5K) on the Ni(110) surface can be easily imaged with a low-temperature STM (see Figure 1). The Xe atom contributes state density to the metal surface near E_F via its unfilled 6s resonance [6] and it appears as a bump of 1.5 Å height in the STM image. Reproducible transfer of the Xe atom between the Ni surface and the W tip achieved by applying voltage pulses to the tip [7] allows us to prepare

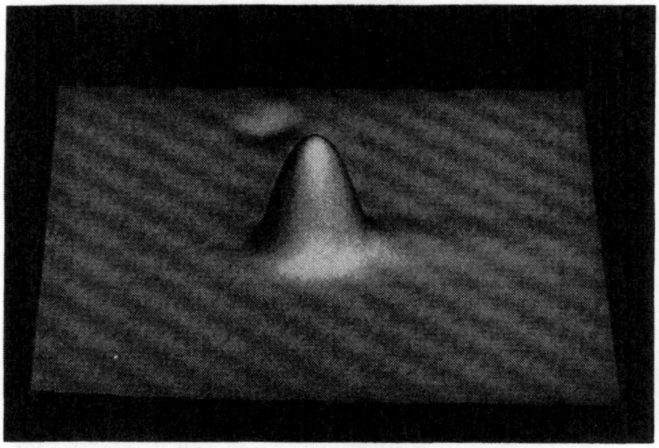

Figure 1. A 50Å×50Å constant current STM image of Xe adsorbed on Ni(110) taken with tip bias +0.01V and a tunneling current of 5×10^{-10} A. Xe appears as a 1.5 Å bump. The rows of the Ni(110) surface are also visible.

an atomically sharp STM tip with a Xe adsorbed at the very end of it. The polarity of the voltage pulse determines the direction of the Xe transfer.

Preferred binding of the Xe atom to low-coordination sites, such as the end of an STM tip, is demonstrated best in an experiment with Fe atoms on the Pt(111) surface shown in Figure 2. In this experiment, a Xe atom is repeatedly transferred between an area of a Pt(111) single crystal surface with three isolated Fe atoms (imaged as 0.9Å bumps) and the end of the tip. Consecutive imaging of this area after each Xe-transfer reveals that the Xe atom only binds to the Fe atom sites showing a large contribution to the local density of states at these sites. Further evidence of binding of the Xe atom to the very end of the tip comes from the enhanced imaging condition that occurs when we transfer the Xe atom from the surface to the tip. (Image in Figure 1 is taken with a Xe-terminated tip).

Several years ago, it was demonstrated that the reversable transfer of a single Xe atom between the Ni surface and the W tip can be used to construct a bistable atomic switch [7]. The "high" conductance of this device which corresponds to the Xe atom adsorbed on the tip exhibits remarkable reproducibility suggesting the importance of Xe's atomic character in determining the conductance value. Another example of the importance of the electronic structure of a single atom on the conductance of a tunnel junction is a recent experiment by Keijsers *et. al* [8] which probes the effect of He atoms in MCB's at low temperatures. The results of this study which is shown in Figure 3 demonstrates that the presence of small amount of He in the tunnel junction at low temperatures results in deviation from the expected exponential dependence of the tunneling current on the electrode spacing. The electrode spacing in the regime of the deviation has been es-

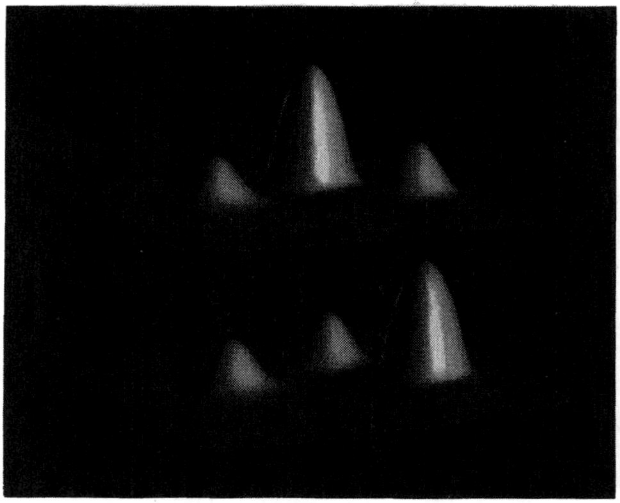

Figure 2. Two STM current images of a 50Å×50Å area of Pt(111) surface with three single Fe atoms and a Xe atom. The Xe prefers to bind to the Fe sites.

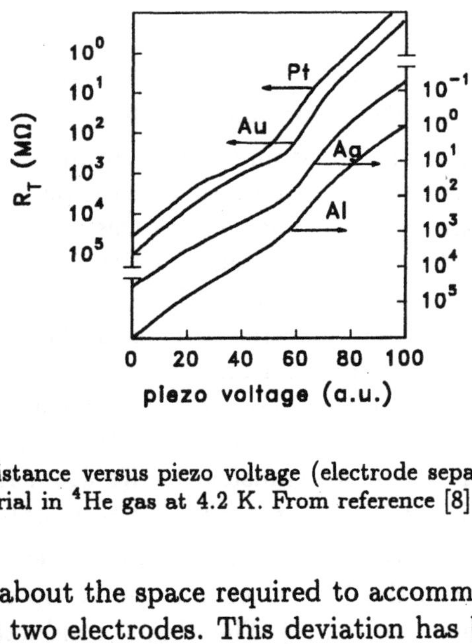

Figure 3. Tunnel resistance versus piezo voltage (electrode separation) for MCB junctions of different material in ^{4}He gas at 4.2 K. From reference [8] with permission.

timated to be just about the space required to accommodate a layer of He atoms between the two electrodes. This deviation has been interpreted as evidence for the reduction of the metal electrodes density of states near E_F caused by the presence of the He 1s orbital, as predicted by Lang [9].

4. Xe Atomic Wires

To measure the resistance of Xe atoms in the STM junction, we first prepare a Xe-terminated tip, by applying a voltage pulse (+1.0V) to the tip to adsorb the Xe atom at the very end of the tip (see inset of Figure 4). We then proceed to measure the resistance (with tip biased at -10 mV) of a single Xe atom coupled to two electrodes by measuring the current I through

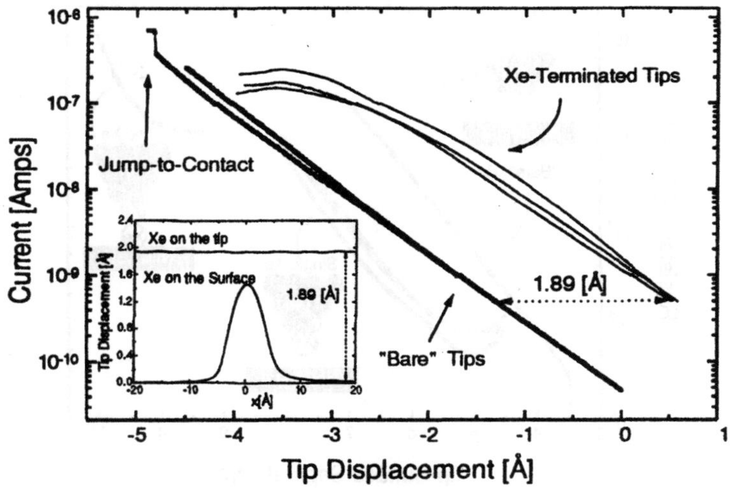

Figure 4. **Current measured through the STM junction (tip bias -10mV) while moving the tip normal to the surface with and without the Xe atom on the tip. Inset shows constant current scans before and after Xe pickup.**

the STM junction while moving the tip toward the Ni surface. Figure 4 shows the measured current for three different Xe-terminated tips, which were prepared from the same polycrystalline W wire by colliding it with the Ni surface or by field emitting from it such that there was a significant change in the microscopic configuration of its outermost atoms, and then terminating the tip with a Xe atom. Also shown in this figure is the data for the "bare" tips which show the expected exponential dependence of the tunneling current I on the spacing between the two electrodes. The Xe-terminated tips, however, show deviation from the exponential behavior of I as the tip approaches the surface, indicating the direct conduction between the tip and the surface through the Xe atom. The offset between the curves for the Xe-terminated tips and the curves for the "bare" tips shown in Figure 4 was determined by the difference in the initial height of the tip above the surface as demonstrated in the inset of Figure 4.

After each measurement with a Xe-terminated tip, we imaged the surface and found it to be undisturbed. We cannot rule out the lateral motion of the Xe atom during our experiments; however, we have confirmed the presence of the Xe atom as the outermost atom on the tip after each experiment by transferring the Xe back to the surface and imaging it. An important point to note is that the linearity of the measured I as a function of the bias voltage has been checked up to biases of up to ± 30mV in all the experiments reported here [10].

We can extract the extra current due to the presence of the Xe atom in

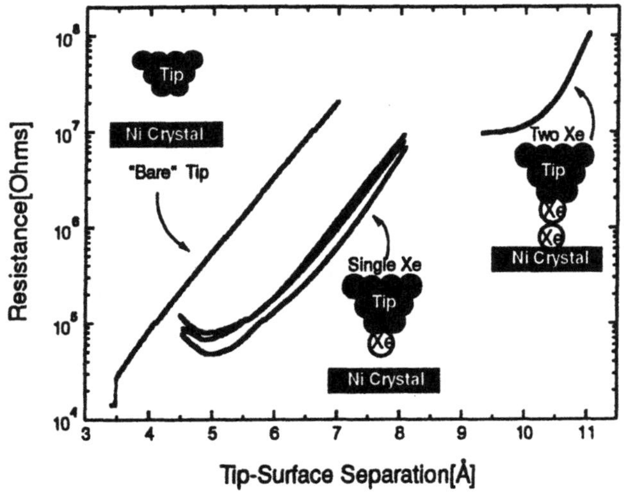

Figure 5. **The measured resistance due to one and two Xe atoms. Also shown in the resistance of a metal-vacuum-metal junction.**

the STM junction by subtracting the I versus D (Displacement) curve for the "bare" tips from that of the Xe-terminated tips. The resulting Xe associated current for one of the tips is shown in Figure 5 as the corresponding resistance plotted as a function of the tip above the surface. Also shown in Figure 5 is the resistance of the "bare" tips that exhibit the characteristic jump-to-contact at $R \simeq 14$ kilohms associated with one-atom metal contacts. To determine tip-surface separation for this figure, we have compared the measured I versus D curves for the "bare" tips to a calculation by Lang[11], which models the STM junction as two jellium surfaces with a single atom adsorbed on one of the surfaces. This calculation shows that an impedance of 2×10^7 ohms for an STM junction corresponds to a tip height above the surface of $\sim$ 7Å. Previously, we found Lang's calculation to be consistent with I versus D measurements taken on this same Ni surface [6]. Using this calibration together with the measured changes in tip height due to the presence of one or two Xe's in the junction, we determine the tip-surface separation.

Binding of the Xe atom to the end of the STM tip gives us the opportunity to measure the resistance of a chain of two Xe atoms in the junction instead of just one. In this measurement, a Xe-terminated tip is centered over a Xe atom adsorbed on the Ni surface before we ramp the tip toward the sample. The current through the STM junction is measured during the ramp and the corresponding resistance is plotted in Figure 5. The surface Xe atom was bound to the same surface site after the experiment as before; however, we cannot rule out lateral motion of the Xe atom during the experiment. Figure 5 shows that the resistance of the STM junction dramatically increases when we have two Xe atom in the junction instead of one. In particular, the leveling off of the resistance at 10 megohms measured for the chain of Xe atoms is 100 times greater than that measured for the single-Xe atom in close contact with both electrodes.

The high resistance of Xe wires as compared to that for one-atom metal links can be understood if we consider the electronic structure of a Xe atom. The Xe atom's state density at the E_F of the electrodes comes almost entirely from the tail of the Xe's broadened 6s resonance which lies 4 volts above E_F. Conduction of the electrons through a Xe atom is therefore determined by the position and broadening of this far-off resonance. A model calculation of tunneling conductance in 3D resonance structures by Kalmeyer and Laughlin [12] shows that the conductance of such structures is strongly dependent on the energy of the resonant state relative to E_F. For the off-resonance case, the conductance is suppressed from the ideal $2e^2/h$ by the ratio of the broadened electronic density at E_F to the resonant energy. In view of that model calculation, we can understand the greater resistance of the single Xe wire based on the lack of strong overlap between the 6s level and states near E_F of the electrodes. Metal atoms adsorbed on metal electrodes have greater state density at E_F and hence exhibit resistances much closer to the ideal 1D channel.

5. Self-consistent Calculations

For a quantitative understanding of the conduction process in our Xe wires, we describe a detailed theoretical model and calculations. Previously [13], our computational procedure was used to determine the resistance of atomic wires consisting of one or more Al atoms, so here we only outline the method of calculation and report the results for the Xe wires. In this procedure, we represent the biased electrodes with the uniform-background (jellium) model [14] and compute the self-consistent electron density distribution for the system in the presence of one or two Xe atoms, whose cores are represented with the pseudopotential of Hamann, Schlüter and Chiang [15]. We calculated the wave functions for propagating states whose energies lie between the E_F's of the two electrodes, so that we could obtain the additional current I that flows due to the presence of the Xe atom or atoms. Dividing the bias voltage by I, we obtained the resistance due to the presence of the atom or atoms, which is directly comparable with the experimental results if the direct tunneling current between the tip and the sample is subtracted from the total I before obtaining the experimental value of the resistance (single Xe case) or when the tunneling current is negligible compared to

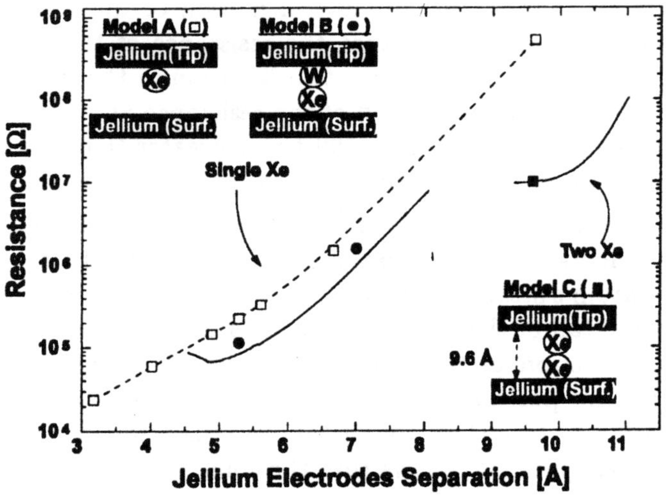

Figure 6. Model calculations (symbols) and the experimental data (same as Figure 5, solid lines). Insets: Model geometries for the calculations.

the total current (two-Xe case). Earlier computations for the single-Xe case were carried out as a part of the work reported by Eigler *et al.* [6] and by Joachime *et al.* [16].

For a single Xe atom in the junction, we considered that the Xe is adsorbed onto one of the electrodes at a distance d and it is a distance s from the other jellium-like electrode (model A, see Figure 6). We calculated the resistance in this case for a number of different s and d values [17]. In Figure 6, we plot the results as a function of $s + d$, which corresponds to the separation between the tip and the Ni surface. The experimental and theoretical resistances shown in Figure 6 are comparable in value but different in their dependence on the separation. The experimental curves indicates a clear change of response when the Xe is close to both electrodes, whereas the theory curve appears nearly exponential over the entire range. In this regime, compression of the Xe atom by the tip may result in the lateral motion of the Xe atom out of the STM junction during the experiment. This possibility is not included in our model calculation and could account for the discrepancy. We also consider a more realistic model of the tip by including an extra metal adatom (W) in the junction (Model B, see Figure 6). As demonstrated in Figure 6 for two tip-surface separation values, this model falls somewhat closer to the experimental curve.

We then modeled the case of the atomic wire consisting of two Xe atoms in between the two jellium electrodes. We calculated the resistance for this wire with the two Xe atoms at their bulk separation (4.4 Å), and with each Xe atom bound to one of the electrodes at the theoretical binding distance (for isolated Xe atom on a jellium surface) of 2.6 Å[18]. In this case, the resistance was 9.4 megohms, in agreement with the experimental data (Figure 6). The striking difference between the resistance of the two-Xe wire and that of the single-Xe case and the one-metal-atom contact, borne out both by the experiment and calculation, can be understood by referring to the Landauer formula [19] for conductance of a single-channel conductor connected to two large contacts: $G = (e^2/\pi\hbar)T$. Here T is the transmission coefficient for the channel and the chemical potential difference for determining the conductance is that measured between points deep in the electrodes. Now consider the transmission properties of two 1D potential wells in series (separated by barriers). If the wells are identical and we consider the conductance at the resonant energy of the well, we expect T to be equal to unity whether we have one or two wells. This situation represents the case of metal atoms with wide resonances centered near the E_F's for the two electrodes [13], where we expect the resistance to be of the order of 12 kilohms and insensitive to the number of atoms in the junction. The case of Xe, however, involves far-off-resonance conduction where T is very small; the T_s for the series of wells, that is, a chain of Xe atoms, can be quite different from that for one well. For example, in our 1D analogy we expect T_2 of the double-well case to be $\sim T_1^2/4$ [20]. Given this relation, it is not surprising that the resistance of the two-Xe wire is 100 times greater than that for a single Xe atom.

6. Conclusion

In summary, we have performed measurements and calculations of the electrical resistance of atomic-size wires consisting of one and two Xe atoms in the gap between an STM tip and a metal surface. We have found a striking change of the resistance by a factor of 100 when we have two Xe atoms instead of one in our STM junction. This difference was accounted for using our detailed fully-self consistent calculations and explained by a far-off-resonance model of the conduction process for Xe atoms. Our results provide the first example of a detailed study in atomic-sized systems where atom-specific information is required to capture the properties of the electrical conduction.

7. Acknowledgment

It is our pleasure to thank C. P. Lutz and R. Landauer for useful discussions. We also thank R. Keijsers and H. van Kempen for providing us with their data.

References

1. C. W. J. Beenaker and H. van Houten, *Solid State Physics*, H. Ehrenreich and D. Turnbull, Eds. (Academic Press, New York, 1991), vol. 44, pp. 1-228 and references therein.
2. B. J. van Wees *et al.*, Phys. Rev. Lett. **60**, 848 (1988)
3. A. Yazdani, D. M. Eigler and N. D. Lang, Science **272**, 1921 (1996).
4. T. N. Todorov and A. P. Sutton, Phys. Rev. Lett. **70**, 2138 (1993), A. M. Bratkovsky, A. P. Sutton, and T. N. Todorov, Phys. Rev. B **52**, 5036 (1995).
5. For example see L. Olesen *et al.*, Phys. Rev. Lett. **72**, 2251 (1994), J. M. Krans and J. M. van Ruitenbeek, Phys. Rev. **50**, 17659 (1994), J. M. Krans *et al.*, Nature **375**, 767 (1995) and references therein.
6. D. M. Eigler, P. S. Weiss, E. K. Schweizer, N. D. Lang, Phys. Rev. Lett. **66**, 1189 (1991).
7. D. M. Eigler, C. P. Lutz and W. E. Rudge, Nature **352**, 600 (1991).
8. R. J. P. Keijsers, J. Voets, O. I. Shkyarevskii, and H. van Kempen, Phys. Rev. Lett. **76**, 1138 (1996).
9. N. D. Lang, Phys. Rev. Lett. **56**, 1164 (1986).
10. We have taken special care to keep the voltage drop across the STM junction constant as we extend the z-piezo toward the surface. Adjustments in the applied voltage become important when the resistance of the junction becomes comparable to the finite input resistance (20 kilohms) of our electrometer.
11. N. D. Lang, Phys. Rev. Lett. **36**, 8173 (1987); J. K. Gimzewski and R. Möller, Phys. Rev. B **36**, 1284 (1987).
12. V. Kalmeyer and R. B. Laughlin, Phys. Rev. B **35**, 9805 (1987).
13. N. D. Lang, Phys. Rev. B **52**, 5335 (1995).
14. See, for example, N. D. Lang, in *Solid State Physics*, F. Seitz, D. Turnbull, H. Ehrenreich, Eds. (Academic Press, New York, 1973), vol. 28, pp. 225-300.
15. D. R. Hamann, M. Schlüter, C. Chiang, Phys. Rev. Lett. **43**, 1494 (1979); G. B. Bachelet, D. R. Hamann, M. Schlüter, Phys. Rev. B **26**, 4199 (1982).
16. C. Joachime *et al.* Europhys. Lett. **20**, 697 (1992).
17. We set $d = 2.6$Å, the equilibrium Xe binding distance [18], except when $s < 2.6$Å where we then set $s = d$ to account for the compression of the Xe atom.
18. N. D. Lang, Phys. Rev. Lett. **46**, 842 (1981).
19. See for example R. Landauer, Phys. Scr. **T42**, 110 (1992) and J. Phys. Condens. Matter **1**, 8099 (1989); Y. Imry in *Directions in Condensed Matter Physics: Memorial Volume in Honor of Shang-keng Ma*, G. Grinstein and G. Mazenko, Eds. (World Scientific, Singapore, 1986), pp. 101-163.
20. M. Büttiker, IBM J. Res. Dev. **32**, 63 (1988); E. O. Kane, in *Tunneling Phenomena in Solids*, E. Burstein and S. Lundqvist, Eds. (Plenum, New York, 1969), pp. 1-11.

CONTACTING MOLECULAR NANOSTRUCTURES.

Molecular Mechanics, Charge Transfer, and Transport Properties

T. A. JUNG, R. R. SCHLITTLER AND J. K. GIMZEWSKI
IBM Research Division
Zurich Research Laboratory
Säumerstrasse 4
8803 Rüschlikon
Switzerland

AND

H. TANG AND C. JOACHIM
CNRS-CEMES
29, rue Jeanne Marvig
31055 Toulouse Cedex
France

Abstract. We present methods for the controlled generation of molecular nanostructures from fullerene and porphyrin-derived molecules. We observe the internal molecular structure of individual molecules, which provides information on the orientation and internal conformation. Molecules and molecular structures produced on various substrates are characterized by scanning tunneling spectroscopy and current distance plots. Molecular mechanics, in response to varying substrate and junction conditions, is explored based on both experimental data and numerical simulations.

1. Introduction

For several decades, the continued downscaling of devices has produced higher density and higher performance semiconductor electronic structures [1]. However, for silicon-based electronic devices to go below dimensions of 100-nm, device manufacturers face various fundamental [2] and economical [3] limitations.

As a possible alternative to silicon technology the integration of molecular components in circuitry has been proposed. This concept, first estab-

C. Joachim and S. Roth (eds.), Atomic and Molecular Wires, 169-177.

lished for donor–σ–acceptor compounds [4], turned out to involve enormous difficulties despite conceptually attractive features. Recently, remarkable progress in the controlled room-temperature assembly of molecular structures [5] has enabled the 'bottom-up' integration of molecules [6]. One of the most attractive features of this approach is that chemical synthesis allows the generation of absolutely identical molecular units, whereas the reproducibility of semiconductor junctions suffers from the decrease in dimensions. In this paper we will study the creation of controlled molecular structures at surfaces together with the characterization of transport perpendicular to the surface using scanning tunneling microscopy (STM) and spectroscopy [7]. It will be shown that molecular assemblies can be designed and assembled and that they serve as well-defined model systems for the study of electronic transport. Furthermore we will illustrate the influence of the interface on the mechanical and electronic properties of a molecule.

2. Molecular Epitaxy of Porphyrins and Fullerenes

The studies reported here have been performed using two different classes of molecules with characteristic and well-defined properties. C_{60}, the most prominent fullerene [8], was chosen because of its electron affinity and spherical shape, Fig. 1a. The rotational symmetric structure of C_{60} gives rise to many properties, for example thermally activated rotation of the molecules in the solid lattice [9]. The strained partial π bonds of the fullerene allow a large electron-deficient [10] delocalized electron system

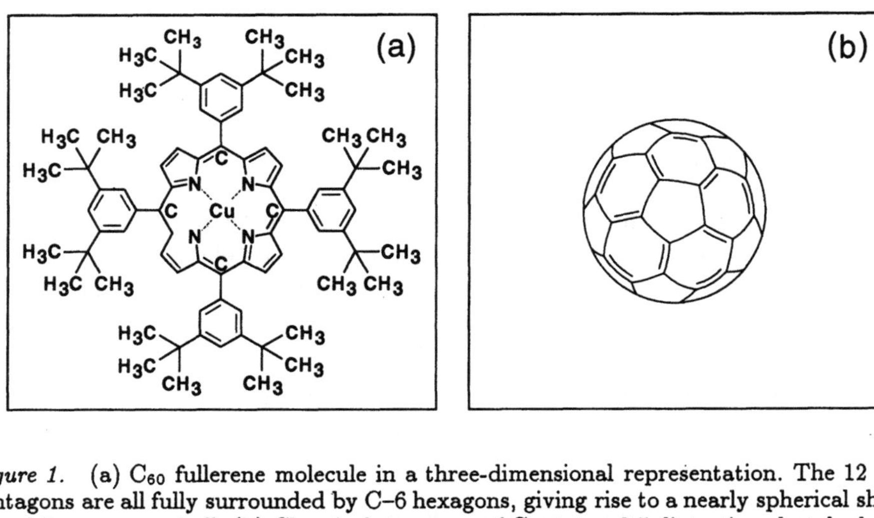

Figure 1. (a) C_{60} fullerene molecule in a three-dimensional representation. The 12 C–5 pentagons are all fully surrounded by C–6 hexagons, giving rise to a nearly spherical shape resembling a soccer ball. (b) Chemical structure of Cu–tetra–3,5 di–tertiary butyl–phenyl porphyrin. Four pyrrole rings are linked around a central metal atom to form a planar delocalized electron system.

that dominates the chemistry [11] and physics [12] of the fullerene. Porphyrins, on the other hand, have many complementary properties. The planar delocalized electron system consists of four pyrrole rings, and the electronic properties of the molecules can be modified by changing the central metal atom or by substitution of H atoms on the outer ring system, Fig. 1b. Porphyrins can be made electronically conductive by *p*-doping and serve as active 'core' units of the catalysing protein of many biochemical reactions. Here we have studied the Cu–tetra–3,5 di–tertiary–butyl–phenyl–porphyrin (Cu–TBPP), which belongs to a class of porphyrins with four phenyl-based substituents. The substituents are semi-flexibly attached to the porphyrin plane as a consequence of sterically inhibited overlap of the porphyrins' and the phenyls' π-electron system. This conformational flexibility has been observed in simulations [5], STM images of adsorbed molecules [13], X-ray crystallography of molecular crystals [14], nuclear magnetic resonance (NMR) data, and isomerization experiments for molecules in solution [15].

Both classes of molecules form epitaxial and quasi-epitaxial layers on many substrates [16, 17]. They have also been characterized by STM and low-energy electron diffraction (LEED). Optimization of the sample-specific growth conditions give arise to the self-assembly of various structures con-

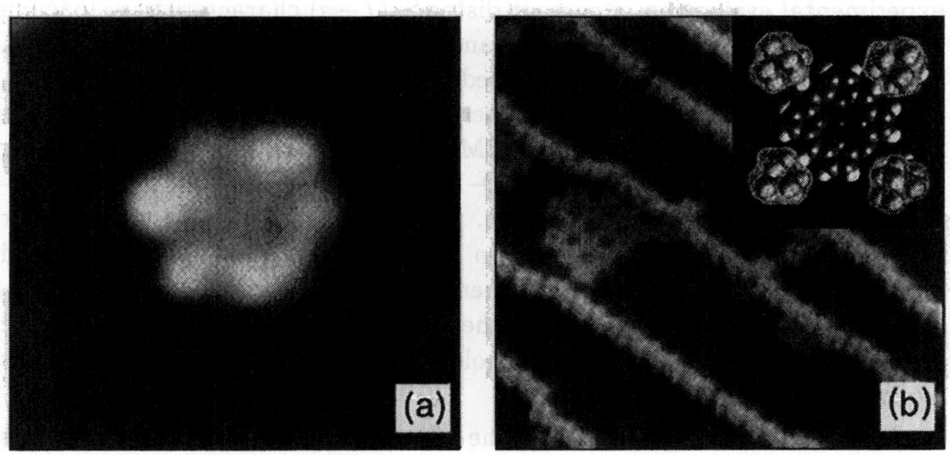

Figure 2. (a) Molecular island of seven C_{60} molecules adsorbed on the Au(110) surface. The 2×1 reconstruction of the Au(110) surface is visible as rows with about 1 Å corrugation. (b) Cu(100) substrate steps decorated with linear arrays of Cu–TBPP. The adsorption site on top of the step edge is favored over the center of terraces for nucleation of an island. Inset: Molecular model showing the protruding substituents responsible for contrast in STM shown as the four light clusters at the outer edges. STM contrast does not correspond to the local density of molecular states as explained in the text.

trolled by kinetic factors and thermodynamics. As an example of the diversity of structures that can be produced, Fig. 2 demonstrates a sixfold island of seven C_{60} fullerenes on Au(110) and an array of porphyrin ('wires') grown by controlled step decoration on Cu(110).

3. Tunneling Transport Characterization

Here we shall consider two distinct cases of electron tunneling transport through a molecule. First, we present C_{60}, whose molecular electronic states are strongly coupled to the substrate. Second, we present two examples of porphyrins which, in contrast, are weakly coupled to the substrate. Here such coupling is hindered by saturated substituents that support the molecule's delocalized electron system above the metal substrate.

The tunneling probability of the STM gap is determined in part by the joint density of states of the tip and sample. Tunneling through such a gap is modified by the presence of a molecule. In spectroscopic experiments, the molecule's electronic states associated with the highest occupied molecular orbitals (HOMO) and the lowest unoccupied molecular orbitals (LUMO) can be identified as distinct features corresponding to the HOMO–LUMO gap energy in $(dI/dV)(V/I)$ vs. V spectroscopy [18]. The tunneling transport through the gap is dominated by the contributions of the HOMO–LUMO tails close to the Fermi level E_F. In a systematic theoretical and experimental evaluation of current distance $(I-s)$ characteristics, Joachim et al. determined the contact point and resistance of a single C_{60} molecule, Fig. 3a [19]. In a further study based on low-voltage I/V characteristics, they showed that, after contact, closer proximity of the tip to the molecule results in a compression of the HOMO–LUMO gap and a corresponding shift and broadening of the HOMO–LUMO states, Fig. 3b [20]. This effect enhances the tunneling current. It is important to note that neither Coulomb charging nor single-electron tunneling effects have been observed for this system, which is a consequence of the close coupling of the C_{60} delocalized electronic system into the metal substrate. After contact, the transport mechanism through the molecule is thus modified predominantly by its mechanical distortion.

In contrast to the case of C_{60}, the STM image contrast of porphyrin is dominated by both topography and the density of states. As can be seen by comparing Fig. 2b with the model presented in the inset, the substituted porphyrin molecule is imaged as four protrusions corresponding to the protruding substituents in square symmetry. The measured apparent height is ~6.5 Å, significantly lower than its height of ~9 Å determined from its molecular structure. This is a consequence of the large HOMO–LUMO gap of the fully saturated tertiary butyl groups. In spite of the

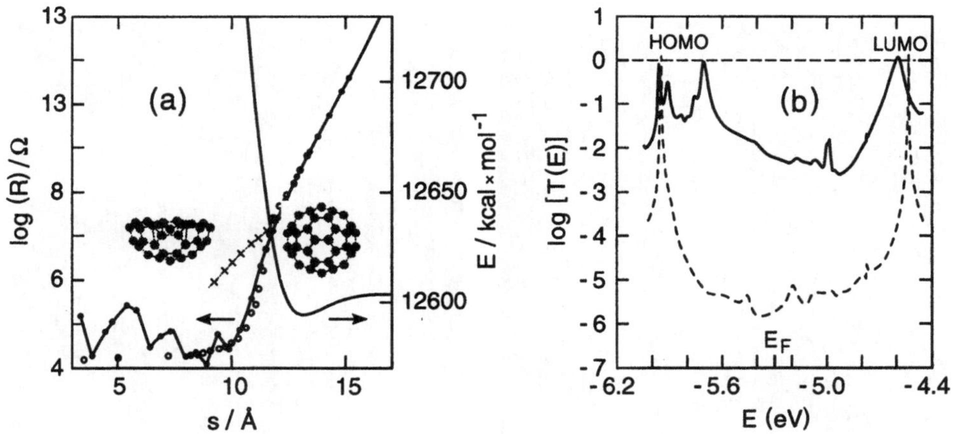

Figure 3. (a) Variation of the STM junction resistance measured across a C_{60} molecule as a function of the distance s between the tip apex and the surface (experiment, open circles; calculation, solid dots). The insets show the optimized structure of the C_{60} molecule for $s = 14$ Å and $s = 7.35$ Å. At the contact point ($\approx$12 Å tip–sample separation), mechanical distortion of the C_{60} molecule begins, giving rise to the changed slope. (b) Variation of the transmission coefficient $T(E)$ through a C_{60} molecule as a function of the electron incident energy calculated with the STM–ESQC (electron scattering quantum chemistry) technique around the HOMO–LUMO gap for $s = 11.4$ 14.4 Å. The HOMO–LUMO tails extended to the Fermi level give rise to conduction that is 2 to 3 orders of magnitude higher due to the compression of the HOMO–LUMO gap in response to mechanical deformation.

small HOMO–LUMO gap associated with the delocalized electron system, the center porphyrin is virtually invisible in the STM experiment. This is a consequence of the weak overlap of the surfaces electronic states with the π system due to the large separation of $\sim$5 Å.

To compare the influence of the separation of the porphyrin ring above the substrate, two porphyrin compounds with different substituents were investigated. They were co-adsorbed onto a Cu(100) substrate at room temperature. The second porphyrin is a modification of Cu–TBPP introduced in Figs. 1a and 2b. Four phenyl substituents are symmetrically attached to the porphyrin as opposed to the di–tertiary–butyl–phenyl substituents of the first molecule. As visible in the model illustrated in Fig. 4a, the shorter supporting legs of Cu-tetra–phenyl–porphyrin (Cu–TPP) result in the porphyrin plane lying closer to the metal surface. The resulting increased π–metal interaction of the porphyrin plane also causes the supporting phenyls to rotate, which further decreases the separation [13]. The resulting overlap of the molecule and surface electronic system gives arise to the observed contrast change. Experimentally, we found that tetraphenyl porphyrin molecules (species 'X' in Fig. 4b) are imaged as rectangular

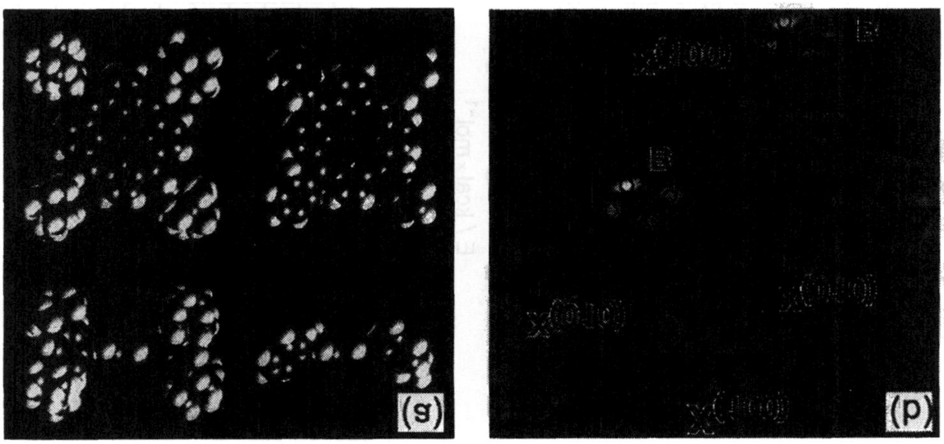

Figure 4. Coevaporation of Cu–TBPP and Cu–TPP shown for contrast and comparison. (a) 3-D models of the two compounds; the porphyrin–metal separation is ≈5.5 Å for Cu–TBPP and less than 2.5 Å for Cu–TPP. As a result of the increased π–metal interaction due to the closer proximity of the π systems with the metallic substrate, the substituents tilt antisymmetrically. (b) Characteristic squarely symmetric protrusions of Cu–TBPP are identified as opposed to two orientations of cross-like features oriented along the 010 and 100 directions (X_{010}, X_{100}). Owing to the different coupling of the molecules to the metallic substrate, opposite contrast in observed for Cu-TPP and Cu–TBPP.

depressions on the Cu(100) substrate, whereas Cu–TBPP (species 'B' in Fig. 4b) shows the characteristic four-lobed protrusions. We note that, because both molecular types are imaged by the same tip apex, the negative contrast of the porphyrin is not associated with a tip effect.

Similar to the case of surface [21] and image states [22] on metals, this contrast inversion is a consequence of the change in the density of electronic states accessible to STM tunneling in combination with the interference of resonant tunneling paths as discussed in [23]. The closer proximity of the porphyrin ring system with the surface metallic states gives arise to some charge transfer due to the mixing of molecular orbitals and surface states. Numerical simulations using the STM–electron scattering quantum chemistry (STM–ESQC) technique [24] are in progress to complement experimental data and to explore the underlying mechanisms.

4. Molecular Bilayers as Model Systems for Charge Transfer

To explore further the electronic and mechanical interaction of molecular species we have prepared a fraction of a monolayer coverage of the nucleophilic Cu–TBPP on the Cu(100) substrate. As an electron-accepting counterpart, a low-submonolayer coverage of C_{60} has been sublimed on top of this sample. Such molecular bilayer systems can serve as an analog to covalently bound donor–σ–acceptor compounds mentioned in the introduction. This supramolecular approach using organic multilayer epitaxy enables the coupling of very different functional units into molecules.

Figure 5 shows an STM image of the prepared multilayer. Image contrast of the multilayer system is slightly weaker than for an individual C_{60} molecule. This sample indicates an alteration of the electronic coupling of the C_{60} to the metallic surface as opposed to the conduction coupling of C_{60} to the metal as described above. It remains to be seen whether Coulomb charging effects, charge transfer, and the shift of electronic and optical properties can be observed for this system or for other donor–acceptor multilayers.

The spectroscopic STM techniques combined with numerical simulations and applied to the well-characterized model systems presented here provide unique insight into the understanding of interfacial molecular properties and molecular electronic transport. Studies of this type have recently been performed for simpler biatomic model systems [25]. The combination of specifically designed molecules with their unique electronic states instead of atoms enables the study of the more complex effects.

This paper demonstrates how molecular properties at surfaces can be probed using STM and how they can be modeled and understood. Both mechanically and electronically, we observe strong interfacial adaptations

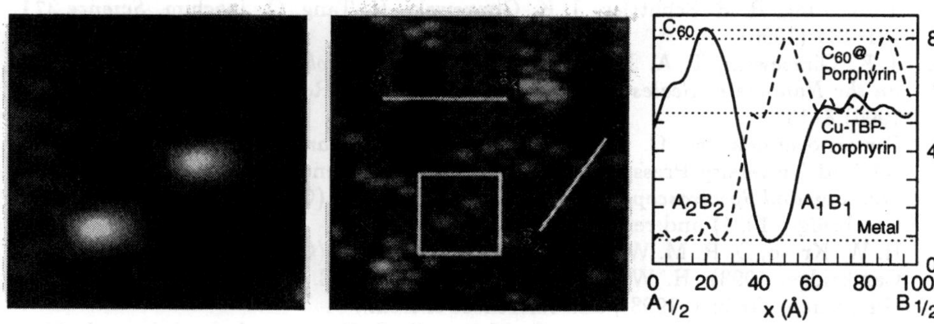

Figure 5. (a) STM image of individual C_{60}. Molecules adsorbed on Cu–TBPP islands grown on Cu(100). The C_{60} molecules can be resolved at the bridge sites between four porphyrin molecules. (b) Overview image with a larger range: Porphyrin islands and the clean substrate can be identified. The area of image (a) has been marked by a rectangle, as well as the cross sections. (c) Cross sections of the data in (b): The apparent STM height of an individual C_{60} adsorbed on top of a clean Cu(100) terrace is slightly higher than the apparent STM height of those C_{60} molecules adsorbed on top of the layer. This reflects the changed transport property of the C_{60} molecule as opposed to that of the multilayer.

of conformation and electronic structures. Technically these effects are important in the context of organic thin films and a variety of applications thereof, as well as in the field of nanotechnology, where it is desirable to observe and reproducibly manipulate and probe nanometer-sized systems. Arrays of complex molecules show promise for the integration of individual units into larger functional systems.

ACKNOWLEDGMENTS

It is a pleasure to thank M. T. Cuberes, P. Guéret, and H. Rohrer for enlightening discussions. We gratefully acknowledge Y. Sakata, K. I. Sugiura, and T. Takami of Osaka University, Japan, for synthesizing and providing the porphyrin molecules. This work was partially supported by grants from the Swiss Federal Office for Education and Science through the ESPRIT basic research project "Processing on the nanometer scale (PRONANO)," No. 8523.

References

1. T. Kaga, M. Ohkura, F. Murai, N. Yokoyama, E. Takeda, *J. Vac. Sci. Technol. B* **13**, 2329–2334 (1995).
2. S. J. Wind, Y. Taur, Y. Mii, D. J. Frank, H.-S. Wong, D. A. Buchanan, S. A. Rishton, J. J. Bucchignano, Y. Lii, K. A. Jenkins, *Mat. Res. Soc. Symp. Proc.* **380**, 179–185 (1995).
3. G. D. Hutcheson and J. D. Hutcheson, *Scientific American*, 54–62 (January 1996).
4. A. Aviram, M. A. Ratner, *Chem. Phys. Lett.* **29**, 277 (1974); A. Aviram, M. J. Freiser, P. E. Seiden, W. R. Young, U.S. Patent 3 953 874, April 27,(1976).
5. T. A. Jung, R. R. Schlittler, J. K. Gimzewski, H. Tang, C. Joachim, *Science* **271**, 181–184 (1996).
6. J. K. Gimzewski, T. A. Jung, R. R. Schlittler, in *Application to Materials Science on the Nanometer Scales*, K. Yoshihara (ed.), (Nat'l Res. Inst. f. Metals, Tsukuba, 1996) 8–13.
7. For an overview see: C. J. Chen, *Introduction to Scanning Tunneling Microscopy* (Oxford University Press, New York, 1993); R. Wiesendanger, *Scanning Probe Microscopy and Spectroscopy: Methods and Applications* (Cambridge University Press, Cambridge, 1994) and references therein.
8. H. W. Kroto, D. R. M. Walton (eds.), *The Fullerenes* (Cambridge University Press, Cambridge, 1993); H. W. Kroto, J. E. Fischer, D. E. Cox (eds.), *The Fullerenes* (Pergamon, Oxford, 1993), and references therein.
9. C. S. Yannoni, R. D. Johnson, G. Meijer, D. S. Bethune, J. R. Salem, *J. Phys. Chem.* **95**, 9–10 (1991).
10. R. C. Haddon, L. E. Brus, K. Raghavachari, *Chem. Phys. Lett.* **125**, 459–464 (1986); S. Modesti, S. Cerasari, P. Rudolf, *Phys. Rev. Lett.* **71**, 2469–2472 (1993).
11. F. Diederich, C. Thilgen, *Science* **271**, 317–323 (1996), and references therein.
12. W. Andreoni (ed.), *The Chemical Physics of Fullerenes 10 (and 5) Years Later*, NATO ASI Series E: Applied Sciences **316** (Kluwer, Dordrecht, 1996), and references therein.
13. T. A. Jung, R. R. Schlittler, J. K. Gimzewski, submitted.
14. H. Song, C. A. Reed, W. R. Scheidt, *J. Amer. Chem. Soc.* **111**, 6865–6866 (1989); B. S. Erler, W. F. Scholz, Y. J. Lee, W. R. Scheidt, C. A. Reed, *J. Amer. Chem.*

Soc. **109**, 2644–2652 (1987); T. K. Miyamoto, N. Sugita, Y. Matsumoto, Y. Sasaki, M. Konno, *Chem. Lett.*, 1695–1698 (1983).
15. S. S. Eaton, G. R. Eaton, *J. Amer. Chem. Soc.* **97**, 3660–3666 (1975); M. J. Crossley, L. D. Field, A. J. Forster, M. M. Harding, S. Sternhell, *J. Amer. Chem. Soc.* **109**, 341–348 (1987); J. W. Dirks, G. Underwood, J. C. Matheson, D. Gust, *J. Org. Chem.* **44**, 2551–2555 (1979).
16. J. K. Gimzewski, in *The Chemical Physics of Fullerenes 10 (and 5) Years Later*, W. Andreoni (ed.), NATO ASI Series E: Applied Sciences **316** (Kluwer, Dordrecht, 1996) 117–136; J. K. Gimzewski, S. Modesti, T. David, R. R. Schlittler, *J. Vac. Sci. Technol. B* **12**, 1942–1946 (1994); T. David, J. K. Gimzewski, D. Purdie, B. Reihl, R. R. Schlittler, *Phys. Rev. B* **50**, 5810–5813 (1994).
17. P. H. Lippel, R. J. Wilson, M. D. Miller, Ch. Wöll, S. Chiang, *Phys. Rev. Lett.* **62**, 171 (1989); R. Strohmaier, C. Ludwig, J. Petersen, B. Gompf, W. Eisenmenger, *J. Vac. Sci. Technol. B* **14**, 1079–1082 (1996); J. C. Buchholz, G. A. Somorjai, *J. Chem. Phys.* **66**, 573–580 (1977).
18. J. K. Gimzewski, S. Modesti, R. R. Schlittler, *Phys. Rev. B* **72**, 1036–1039 (1994).
19. C. Joachim, J. K. Gimzewski, R. R. Schlittler, C. Chavy, *Phys. Rev. Lett.* **74**, 2102–2105 (1995).
20. C. Joachim, J. K. Gimzewski, *Europhys. Lett.* **30**, 409–414 (1995).
21. Y. W. Mo, F. J. Himpsel, *Phys. Rev. B* **50**, 7868 (1994).
22. T. Jung, Y. W. Mo, F. J. Himpsel, *Phys. Rev. Lett.* **74**, 1641–1644 (1995).
23. A. J. Fisher, P. E. Blöchl, in *Computations for the Nano-Scale*, P. E. Blöchl, C. Joachim, A. J. Fisher (eds.), NATO ASI Series E: Applied Sciences **240** (Kluwer, Dordrecht, 1993), 185–197.
24. P. Sautet, C. Joachim, *Chem. Phys. Lett.* **185**, 23–30 (1991); P. Sautet, C. Joachim, *Surf. Sci.* **271**, 387–394, (1992); C. Chavy, C. Joachim, A. Altibelli, *Chem. Phys. Lett.* **214**, 569–575 (1993).
25. A. Yazdani, D. M. Eigler, N. D. Lang, *Science* **272**, 1921–1924 (1996).

Computation of electrostatic fields around carbon tubules biased by an STM junction

Ch. GIRARD, M. DEVEL AND X. BOUJU
Laboratoire de Physique Moléculaire URA CNRS 772
Université de Franche-Comté, 25030 Besançon, France

AND

P.A. GRAVIL
School of Physics, University of Bath, Bath BA2 7AK, UK

Abstract. When a carbon tubule is electrically biased by an STM junction, an electrostatic field confined at the nanometer scale arises in a region located around and inside the fullerene molecule. In this contribution, we present a semi-empirical approach of this problem within the framework of the field propagator method. Starting from the knowledge of the field distribution associated with a highly symmetrical reference system, we derive the self-consistent field distribution for the whole system by introducing a generalized field-propagator obtained by solving a sequence of Dyson's equations numerically. We then use an appropriate renormalization procedure to avoid the usual divergence occurring at the immediate proximity of the electronic structure. Within the model, the fullerene molecule is treated as a rigid cluster of **n** anisotropic polarizable systems located at the position of the carbon atoms. The application of this numerical scheme for long carbon tubules in STM configurations is discussed.

Key Words: electrostatic field, carbon tubule, STM junction, nanoscale physical processes

1. Introduction

The precise description of the electrostatic field resulting from the self-consistent interactions of adsorbed species trapped in an STM junction, has become mandatory to deal with a growing class of phenomena occurring at the nanometer scale [1]. Recently, several theoretical schemes have been developed in this context. For example, important issues concerning the field induced atomic transfer mechanisms have been discussed by Tsong and compared with basic physical processes occurring in field-ion microscopy (FIM) [2 , 3 , 4 , 5]. Subsequently, the problem of the transfer of a single atom trapped inside a metal-vacuum-metal junction was

C. Joachim and S. Roth (eds.), *Atomic and Molecular Wires*, 179-192.

analyzed theoretically by Lang within the density functional formalism [6 , 7]. More recently, an important breakthrough was achieved by Hirose and Tsukada who proposed a first principle calculation of the electronic structure for an STM junction under strong field and current [8 , 9]. In the case of the Na–vacuum–Na junction, this methodology based on a microscopic description of the electrodes has been able to address the challenging problem of atom extraction by STM.

In the presence of long fullerene nanotubes, the rather large number of valence electrons per fullerene makes it difficult to determine the electric response of the molecule from *ab initio* calculations. As a consequence, one has to consider developing simpler approaches. For example, two different semi–empirical schemes based on a reduced number of physical parameters, have been used to investigate the electrical response of a single C_{60} molecule in the context of physical adsorption near metallic [10] and ionic substrates [11]. In these studies [10,11] the net polarization induced on the C_{60} was approximated by the polarization of 60 coupled dipoles, characterized by a set of anisotropic polarizabilities, centered on the atomic equilibrium positions of each carbon atoms. The *in-situ* polarizabilities of each carbon atom composing the C_{60} network were deduced from the polarizability of graphite by applying a simple spherical to cartesian tranformation of the coordinates.

In this paper, we generalize this procedure to deal with conjugated carbon networks of any local curvature such as capsules or carbon tubules of arbitrary lengths. The guiding principle of our method is based on a combination of macroscopic and microscopic descriptions: starting from the known solution of a bare reference system (BRS), submitted to an external electric potential, the coupling with adsorbates can be accounted for by introducing a set of anisotropic polarizabilities for each atom of the adsorbed molecules [12]. The final field distribution may then be computed self-consistently by introducing a generalized field-propagator obtained by solving a sequence of Dyson's equations numerically. The ability of this procedure to deal with large molecules or molecular islands is a fundamental advantage when studying the physical interaction between biased STM tips and very long carbon tubules. Such an algorithm avoids the numerical inversion of a very large matrix ($\mathbf{3n} \times \mathbf{3n}$) where $\mathbf{n}$ is the total number of carbon atoms composing the fullerene) by reducing the self-consistent problem into a succession of small (3×3) *matrix* inversions. This iterative scheme is therefore relevant for all studies concerned with a great number of degrees of freedom because it is much less demanding in terms of computer time. It is particularly well-suited for considering low-symmetry systems for which first–principle calculations become rapidly intractable.

In order to reach this goal, we will adapt in sections 2 and 3, the electrostatic field propagator method (EFP) to tackle some configurations of current interest in the STM technology (molecular islands trapped in STM junctions, atomic wires connected to macroscopic pads, carbon tubules, ...). As a first step, we will derive the set of self–consistent fields on each atomic site. The whole field distribution will then be deduced from these data by introducing a renormalisation procedure removing the usual divergence occurring at the immediate proximity of the electronic structures. In section 4, we will present our results about the evolution of the field patterns occurring around, as well as, inside single shell carbon tubules located in a biased STM junction.

2. Basic equations

2.1. The microscopic Poisson's equation

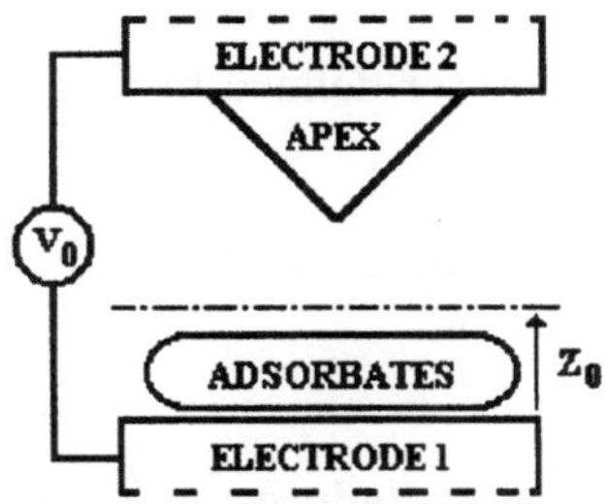

Fig.1: Geometry of the biased tip-sample junction used for the local field calculations. Z_0 defines an observation plane parallel to the sample (electrode 1).

In this section, we briefly present the main features of our theoretical framework. Let us consider a molecule, characterized by its charge density $\rho(\mathbf{r})$, and located in the gap region formed by two metallic electrodes (c.f. Fig. 1). The Poisson's equation associated with this system reads

$$\Delta\phi(\mathbf{r}) = -4\pi\rho(\mathbf{r}). \tag{1}$$

We then express the molecular charge density in term of the non–local polarization $\mathbf{P}(\mathbf{r})$ of the molecule

$$\rho(\mathbf{r}) = -\nabla\cdot\mathbf{P}(\mathbf{r}), \tag{2}$$

and, in a second step, we introduce the self–consistent electric field $\mathbf{E}(\mathbf{r})$ in the gap. This leads to

$$\nabla\cdot\mathbf{E}(\mathbf{r}) = -4\pi\nabla\cdot\mathbf{P}(\mathbf{r}). \tag{3}$$

If we assume a linear relation between the self–consistent field $\mathbf{E}(\mathbf{r})$ and the molecular polarization, we have

$$\mathbf{P}(\mathbf{r}) = \chi(\mathbf{r})\cdot\mathbf{E}(\mathbf{r}), \tag{4}$$

where, in the framework of the linear quantum response theory, $\chi(\mathbf{r})$ is a quadratic function of the matrix elements $\mathbf{P}_{n,m}(\mathbf{r})$ of the polarization operator of the molecule

$$\chi_{\alpha,\beta}(\mathbf{r}) = \sum_{n,m}\sigma_{m,m}\{\frac{P^{\alpha}_{n,m}(\mathbf{r})P^{\beta}_{m,n}(\mathbf{r})}{E_m - E_n} + \frac{P^{\beta}_{n,m}(\mathbf{r})P^{\alpha}_{m,n}(\mathbf{r})}{E_m - E_n}\}. \tag{5}$$

Substitution of Eq. (4) into Eq. (3), gives the usual homogeneous relation

$$\nabla\cdot\mathbf{E}(\mathbf{r}) = -4\pi\nabla_{\mathbf{r}}\cdot[\chi(\mathbf{r})\cdot\mathbf{E}(\mathbf{r})]. \tag{6}$$

Solving this differential equation in the presence of an intricate metallic surroundings calls for appropriate numerical schemes. In the next section we will show how the Electrostatic Field Propagators (EFP) method allows to deal efficiently with this kind of problem.

2.2. Integral representation and generalized propagator

It is straightforward to rewrite Eq. (6) as follows

$$\mathcal{O}(\mathbf{r}) \cdot \mathbf{E}(\mathbf{r}) = 0, \tag{7}$$

where $\mathcal{O}(\mathbf{r})$ is a differential operator defined by

$$\mathcal{O}(\mathbf{r}) = \nabla_{\mathbf{r}} \cdot [\mathbf{I} + 4\pi\chi(\mathbf{r})]. \tag{8}$$

In addition, one can write that the zero order field $\mathbf{E}_0(\mathbf{r})$, i.e. in the absence of molecule, obeys Laplace's equation

$$\nabla_{\mathbf{r}} \cdot \mathbf{E}_0(\mathbf{r}) = 0. \tag{9}$$

From Eqs. (7),(8) and (9), we may deduce the following relation

$$\mathcal{O}(\mathbf{r}) \cdot \mathbf{E}_0(\mathbf{r}) - 4\pi\nabla_{\mathbf{r}} \cdot \chi(\mathbf{r}) \cdot \mathbf{E}_0(\mathbf{r}) = \mathcal{O}(\mathbf{r}) \cdot \mathbf{E}(\mathbf{r}). \tag{10}$$

This new relation allows then to separate the self–consistent field $\mathbf{E}(\mathbf{r})$ from the excitation field $\mathbf{E}_0(\mathbf{r})$, yielding

$$\mathbf{E}(\mathbf{r}) = \mathbf{E}_0(\mathbf{r}) - 4\pi\mathcal{O}^{-1}(\mathbf{r}) \cdot \nabla_{\mathbf{r}} \cdot \chi(\mathbf{r}) \cdot \mathbf{E}_0(\mathbf{r}). \tag{11}$$

Finally, in order to transform this last expression into an integral equation, we introduce a "delta–function" in its right hand side

$$\mathbf{E}(\mathbf{r}) = \int [\mathbf{E}_0(\mathbf{r}') - 4\pi\mathcal{O}^{-1}(\mathbf{r}') \cdot \nabla_{\mathbf{r}'} \cdot \chi(\mathbf{r}') \cdot \mathbf{E}_0(\mathbf{r}')]\delta(\mathbf{r} - \mathbf{r}')d\mathbf{r}'. \tag{12}$$

Now, to go further, we have to introduce the electrostatic field propagators associated with the bare reference system (surroundings) and the entire system (molecule + surroundings) respectively. These two dyadic tensors verify the following differential equations

$$\nabla_{\mathbf{r}} \cdot \mathbf{S}_0(\mathbf{r}, \mathbf{r}') = -4\pi\nabla_{\mathbf{r}}\delta(\mathbf{r} - \mathbf{r}'), \tag{13}$$

and

$$\mathcal{O}(\mathbf{r}) \cdot \mathcal{S}(\mathbf{r}, \mathbf{r}') = -4\pi\nabla_{\mathbf{r}}\delta(\mathbf{r} - \mathbf{r}'). \tag{14}$$

From this last relation, it is possible to express the inverse operator $\mathcal{O}^{-1}(\mathbf{r})$ in terms of $\mathcal{S}(\mathbf{r}, \mathbf{r}')$

$$\mathcal{S}(\mathbf{r}, \mathbf{r}') = -4\pi\mathcal{O}^{-1}(\mathbf{r}) \cdot \nabla_{\mathbf{r}}\delta(\mathbf{r} - \mathbf{r}'). \tag{15}$$

After back–substitution of Eq. (15) into Eq. (12), we obtain an explicit integral equation governing the entire electrostatic field distribution around, as well as, inside the molecular system

$$\mathbf{E}(\mathbf{r}) = \int \mathcal{K}(\mathbf{r}, \mathbf{r}') \cdot \mathbf{E}_0(\mathbf{r}')d\mathbf{r}', \tag{16}$$

with

$$\mathcal{K}(\mathbf{r},\mathbf{r}') = \delta(\mathbf{r}-\mathbf{r}') + \mathcal{S}(\mathbf{r},\mathbf{r}') \cdot \chi(\mathbf{r}'). \tag{17}$$

This equation defines the generalized propagator of the system. This quantity contains all the informations on the system and can be derived numerically by using the remarkable properties of Dyson's equations. In that scheme, the dyadic $\mathcal{S}(\mathbf{r},\mathbf{r}')$ can be related to the surrounding's propagator $\mathbf{S_0}(\mathbf{r},\mathbf{r}')$ by the implicit integral equation

$$\mathcal{S}(\mathbf{r},\mathbf{r}') = \mathbf{S_0}(\mathbf{r},\mathbf{r}') + \int \mathbf{S_0}(\mathbf{r},\mathbf{r}'') \cdot \chi(\mathbf{r}'') \cdot \mathcal{S}(\mathbf{r}'',\mathbf{r}')d\mathbf{r}'', \tag{18}$$

where the integral runs over the volume occupied by the molecule. The influence of the metallic surroundings on which the molecule is adsorbed is implicitly accounted for by $\mathbf{S_0}(\mathbf{r},\mathbf{r}')$. For a pair of plane metallic electrodes, this dyadic is described in [12]. As already discussed in our previous published works [10,12] such a development of real space approaches for both low–symmetry and low–dimensional systems is well–suited to cope with complex geometry. A numerical procedure derived from this methodology will be presented in the following section.

3. Numerical procedure

In this section, we propose a numerical scheme to solve both Eqs. (16) and (18). First, let us introduce a model distribution of molecular susceptibility $\chi(\mathbf{r})$ which reproduces the atomic array of the molecule. Even though more general model densities can be used, we choose here a simple representation of atom-centered spherical Gaussians

$$\chi(\mathbf{r}) = \sum_i \alpha_\mathbf{i}(\mathbf{r}-\mathbf{r_i}), \tag{19}$$

where the coefficients $\alpha_\mathbf{i}(\mathbf{r}-\mathbf{r_i})$ are nonlocal atomic polarizabilies and $\mathbf{r_i}$ denotes a particular atomic site. Further computations require knowledge of the explicit forms of $\alpha_\mathbf{i}(\mathbf{r}-\mathbf{r_i})$, which in principle can be evaluated from the atomic wave functions. To bring out the physical mechanism associated with the divergence occurring at the immediate proximity of the electronic molecular structures we will assume that $\alpha_\mathbf{i}(\mathbf{r}-\mathbf{r_i})$ have the form [13 , 14]

$$\alpha_\mathbf{i}(\mathbf{r}) = \alpha_\mathbf{i} f(r), \tag{20}$$

where $f(r)$ is a Gaussian normalized to the unity

$$f(r) = \frac{1}{\pi^{3/2}\eta^3}\exp(-r^2/\eta^2). \tag{21}$$

In Eq. (20) $\alpha_\mathbf{i}$ defines the static *in–situ* polarizability of the ith atom and in Eq. (21) η is a parameter related to the size of atoms. In the two next subsections, we will use this representation to describe the self–consistent field at each atomic site and, then to propagate this field outside atomic structures.

3.1. Solving the self-consistent field for each atomic site

In this section, we show that it is worthwhile to start the numerical procedure with the computation of the self-consistent fields $\{\mathbf{E}(\mathbf{r_i})\}$ occurring on the atomic array. To achieve this goal, Eqs. (16) and (18) will be solved by discretization in the real space by exploiting the natural grid $\{\mathbf{r_i}\}$ provided by the atoms themselves. This leads to

$$\mathbf{E}(\mathbf{r_i}) = \mathbf{E}_0(\mathbf{r_i}) + \int_v \sum_k \mathcal{S}(\mathbf{r_i}, \mathbf{r}') \cdot \alpha_\mathbf{k} \cdot \mathbf{E}_0(\mathbf{r}') f(|\mathbf{r}' - \mathbf{r_k}|) d\mathbf{r}', \tag{22}$$

and

$$\mathcal{S}(\mathbf{r_i}, \mathbf{r_j}) = \mathbf{S}_0(\mathbf{r_i}, \mathbf{r_j}) + \int_v \sum_k \mathbf{S}_0(\mathbf{r_i}, \mathbf{r}') \cdot \alpha_\mathbf{k} \cdot \mathcal{S}(\mathbf{r}', \mathbf{r_j}) f(|\mathbf{r}' - \mathbf{r_k}|) d\mathbf{r}'. \tag{23}$$

At this stage, we can introduce two licit hypotheses. First, we will assume that the decay radius η_k of the Gaussian representation are chosen such that the tails do not spill over the boundaries of the unit atomic cells; second, we will consider that the spatial variations of $\mathbf{E}_0(\mathbf{r}')$, $\mathbf{S}_0(\mathbf{r_i}, \mathbf{r}')$, and $\mathcal{S}(\mathbf{r_i}, \mathbf{r}')$ may be neglected when the position vector $\mathbf{r}'$ runs over an atomic cell. As the functions $f(|\mathbf{r}' - \mathbf{r_k}|)$ are normalized to unity, Eq. (22) then becomes

$$\mathbf{E}(\mathbf{r_i}) = \mathbf{E}_0(\mathbf{r_i}) + \sum_k \mathcal{S}(\mathbf{r_i}, \mathbf{r_k}) \cdot \alpha_\mathbf{k} \cdot \mathbf{E}_0(\mathbf{r_k}), \tag{24}$$

and

$$\mathcal{S}(\mathbf{r_i}, \mathbf{r_j}) = \mathbf{S}_0(\mathbf{r_i}, \mathbf{r_j}) + \sum_k \mathbf{S}_0(\mathbf{r_i}, \mathbf{r_k}) \cdot \alpha_\mathbf{k} \cdot \mathcal{S}(\mathbf{r_k}, \mathbf{r_j}). \tag{25}$$

Using shorthand notations, Eq. (24) may be rewritten as a simple matricial product

$$\mathcal{E} = \mathcal{Q} \bullet \mathcal{E}_0, \tag{26}$$

where the symbol $\bullet$ indicates a total contraction on both Cartesian indexes and positions of discretized elements. Moreover $\mathcal{E}$ and $\mathcal{E}_0$ are two supervectors defined by

$$\mathcal{E} = \{\mathbf{E}(\mathbf{r_1}); \mathbf{E}(\mathbf{r_2}); .....; \mathbf{E}(\mathbf{r_n})\}, \tag{27}$$

and

$$\mathcal{E}_0 = \{\mathbf{E}_0(\mathbf{r_1}); \mathbf{E}_0(\mathbf{r_2}); .....; \mathbf{E}_0(\mathbf{r_n})\}. \tag{28}$$

In Eq. (26), the $(3\mathbf{n}\times 3\mathbf{n})$ matrix $\mathcal{Q}$ is built from the different components taken by the electrostatic field propagator $\mathcal{S}$ for all possible couples of position vectors $\{\mathbf{r_i}, \mathbf{r_j}\}$

$$\mathcal{Q} = \begin{pmatrix} \mathbf{I} + \mathcal{S}(\mathbf{r_1}, \mathbf{r_1}) \cdot \alpha_1 & \mathcal{S}(\mathbf{r_1}, \mathbf{r_2}) \cdot \alpha_2 & \cdots & \mathcal{S}(\mathbf{r_1}, \mathbf{r_n}) \cdot \alpha_n \\ \mathcal{S}(\mathbf{r_2}, \mathbf{r_1}) \cdot \alpha_1 & \mathbf{I} + \mathcal{S}(\mathbf{r_2}, \mathbf{r_2}) \cdot \alpha_2 & \cdots & \mathcal{S}(\mathbf{r_2}, \mathbf{r_n}) \cdot \alpha_n \\ \vdots & \vdots & \ddots & \vdots \\ \mathcal{S}(\mathbf{r_n}, \mathbf{r_1}) \cdot \alpha_1 & \mathcal{S}(\mathbf{r_n}, \mathbf{r_2}) \cdot \alpha_2 & \cdots & \mathbf{I} + \mathcal{S}(\mathbf{r_n}, \mathbf{r_n}) \cdot \alpha_n \end{pmatrix}. \tag{29}$$

A computerized calculation of this matrix requires solving the discrete representation of Dyson's equation (25).

3.2. *Solving Dyson's equation*

In this part the electrostatic field propagator $\mathcal{S}$ of an arbitrary reference system (here the junction) covered by a finite number of polyatomic molecules is derived from a sequence of Dyson's equations. For a large number of interacting particles, i.e. all atoms composing the adsorbed molecules, we find it is worthwhile to derive the electrostatic field propagator by using a fully self-consistent iterative scheme. In order to illustrate this procedure we consider a system composed of (**n**) atoms interacting with a complex surrounding [15]. The field susceptibility associated with the first atom is found by application of Dyson's equation

$$\mathbf{S_1}(\mathbf{r},\mathbf{r}') = \mathbf{S_0}(\mathbf{r},\mathbf{r}') + \mathbf{S_0}(\mathbf{r},\mathbf{r_1}) \cdot \alpha_1 \cdot \mathbf{A}_1 \cdot \mathbf{S_0}(\mathbf{r_1},\mathbf{r}'). \tag{30}$$

In this equation, the index (1) refers to the first atom of the adsorbate and the (3×3) tensor $\mathbf{A}_1$ is the result of a small matrix inversion

$$\mathbf{A}_1 = [\mathbf{I} - \mathbf{S_0}(\mathbf{r_1},\mathbf{r_1}) \cdot \alpha_1]^{-1}. \tag{31}$$

In a similar way, the field susceptibility labelled $\mathbf{S_2}$ associated with the two first atoms of the molecular aggregate can be deduced from the knowledge of $\mathbf{S_1}$. This leads to

$$\mathbf{S_2}(\mathbf{r},\mathbf{r}') = \mathbf{S_1}(\mathbf{r},\mathbf{r}') + \mathbf{S_1}(\mathbf{r},\mathbf{r_2}) \cdot \alpha_2 \cdot \mathbf{A}_2 . \mathbf{S_1}(\mathbf{r_2},\mathbf{r}'). \tag{32}$$

At this stage the matrix $\mathbf{A}_2$ is built by replacing $\mathbf{S_0}$ by $\mathbf{S_1}$ in the previous iteration. By repeating this procedure, it is thus possible to derive the field susceptibility $\mathbf{S_n}$ of a discrete system formed of **n** atoms from the field-susceptibility $\mathbf{S_{n-1}}$ of a system composed of (**n**-1) atoms. This leads to

$$\mathbf{S_n}(\mathbf{r},\mathbf{r}') = \mathbf{S_{n-1}}(\mathbf{r},\mathbf{r}') + \mathbf{S_{n-1}}(\mathbf{r},\mathbf{r_n}) \cdot \alpha_\mathbf{n} \cdot \mathbf{A}_n \cdot \mathbf{S_{n-1}}(\mathbf{r_n},\mathbf{r}') = \mathcal{S}(\mathbf{r},\mathbf{r}'). \tag{33}$$

Such an algorithm avoids the numerical inversion of a very large matrix $(3\mathbf{n} \times 3\mathbf{n})$. It reduces the self-consistent problem to a succession of small (3×3) *matrix* inversions.

3.3. *Continuous description of the field inside and outside the molecular structure*

In the previous subsection, we have described a numerical procedure to compute the field vectors $\mathbf{E}(\mathbf{r_k})$ on each atomic site. In order to extend this information throughout the molecular structure, it is more efficient to start with the implicit integral equation associated with the self–consistent field

$$\mathbf{E}(\mathbf{r}) = \mathbf{E}_0(\mathbf{r}) + \int_v \mathbf{S_0}(\mathbf{r},\mathbf{r}') \cdot \chi(\mathbf{r}') \cdot \mathbf{E}(\mathbf{r}')d\mathbf{r}'. \tag{34}$$

At this stage, we will use the same two restrictive hypotheses already introduced to derive Eqs. (24) and (25). First, the self–consistent field given by the first computing round (cf. Eqs. (26) and (27)) varies slowly over the scale of a given atom of the molecule. Second, we will assume that such spatial variations may be also neglected when performing the integration on the $\mathbf{r}'$ vector entering the propagator $\mathbf{S_0}(\mathbf{r}, \mathbf{r}')$ associated to the bare reference system. These two assumptions lead to

$$\mathbf{E}(\mathbf{r}) = \mathbf{E}_0(\mathbf{r}) + \sum_k \mathbf{S_0}(\mathbf{r}, \mathbf{r_k}) \cdot \int_{v_k} \alpha_{\mathbf{k}}(\mathbf{r}')d\mathbf{r}' \cdot \mathbf{E}(\mathbf{r}_k). \tag{35}$$

In this expression, the integral domains v_k run on the volume occupied by each atomic site except when the modulus of the observation vector $\mathbf{r}$ enters the electronic clouds. In this case, the integration domains around the kth atom must be limited to a sphere of radius $|\mathbf{r} - \mathbf{r_k}|$. After some algebraic manipulations one obtains

$$\mathbf{E}(\mathbf{r}) = \mathbf{E}_0(\mathbf{r}) + \sum_k \mathcal{H}(R_k)\mathbf{S_0}(\mathbf{r}, \mathbf{r_k}) \cdot \alpha_{\mathbf{k}} \cdot \mathbf{E}(\mathbf{r}_k), \tag{36}$$

where the scalar function $\mathcal{H}(R_k)$ is defined by

$$\mathcal{H}(R_k) = erf(\frac{|\mathbf{r} - \mathbf{r_k}|}{\eta}) - \frac{2|\mathbf{r} - \mathbf{r_k}|}{\eta\pi^{1/2}} \exp(-\frac{|\mathbf{r} - \mathbf{r_k}|^2}{\eta^2}). \tag{37}$$

4. Field distributions calculations

In this section, we apply our formalism to local field calculations around large adsorbates located inside an STM junction. We have chosen fullerene molecules because of their current experimental interest in STM [16 , 17 , 18 , 19]. The *in-situ* polarizabilities of each carbon atom composing the fullerene network were deduced from the polarizability of graphite by applying a simple spherical to cartesian tranformation of the coordinates. For a given atom, a local cartesian frame is built from the normal to the plane of the three nearest neighbours, and from two independent directions chosen in this plane. In this local frame, the polarizability tensor is always diagonal, with $\alpha_{\parallel}^{carb}$= 1.57 Å^3, and $\alpha_{\perp}^{carb}$= 0.865 Å^3, so that the trace of the polarizability tensor is equal to three times the isotropic polarizability. The atomic polarizability tensor is then computed in the absolute frame of the junction by a simple cartesian frame transformation [12].

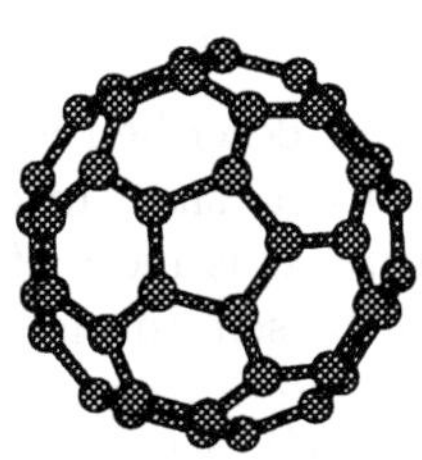

Fig.2: Top view of the C_{60} molecule in the STM junction. This orientation has been chosen to perform the simulation described in Fig. 3.

Note that in our numerical scheme, the presence of a metallic cluster schematizing the tip–apex of the STM, can be accounted for by introducing two perturbations in the gap [12]

$$\chi(\mathbf{r}) = \chi_{apex}(\mathbf{r}) + \chi_{ful}(\mathbf{r}), \tag{38}$$

where each contribution is associated with the apex and with the fullerene respectively.

4.1. Field map around a single C_{60}

As a reference for later comparison, we first focus our attention on a single C_{60} adsorbed on a gold surface (electrode 1). For ideal C_{60} molecules, the spatial arrangement of carbon atoms can be found in [20]. In this preliminary study, the calculation is performed in the absence of any tip–apex so that the upper electrode is just a plane copper slab.

The results gathered in Fig. 3, have been obtained by keeping the metal to metal distance D and the applied voltage V_0 at a constant level (D = 18 Å and V_0 = 2V) and using the same computational window of size (12 Å× 12 Å) centered around the C_{60}. The only difference between the two images of Fig. 3 is the distance Z_0 between the observation plane and the lower electrode (see Fig. 2). In this application, the zero order field (i.e., in the absence of any molecule) reaches 0.111 V/Å and the C_{60} orientation is chosen with a five–membered ring on the top of the molecule. When the calculation is performed at the immediate proximity of the molecule, for example at only 1.5 Å of the carbon centers (cf. first field map of Fig. 3), the field distribution $|E(X,Y)|$ reveals the atomic structure of the last carbon layer. More precisely, the atoms that compose the top five–membered ring localize and confine the field to form five atomic spots (of about 0.32 V/Å in magnitude). This effect is enhanced by anisotropy effects related to the local deformation of carbon orbitals. Moreover, when we decrease the observation plane position so as to pass through the center of the molecule, by lowering the initial observation plane by about 5 Å, we observe a drastic change in the field map (cf. second field map presented in Fig. 3). When comparing these two maps, the most striking effect is the field depression obtained inside the carbon cage. In the configuration where the applied field reaches about 0.111 V/Å, the field reduction between the applied voltage and the local field ΔE varies between -0.05 V/Å and -0.06 V/Å around the center of the cage. The physical origin of the significant magnitude of ΔE is mainly the screening effect produced by the collective polarization of the C_{60} carbon atoms. Moreover, we observe also a strong confinement of the field lines around the carbon atoms that compose the graphitic shell. In this very localized region, the maximum field strength reaches about 0.5 V/Å.

To control the stability of our model, we lifted gradually the observation plane, by adding up to 8 Å to this last configuration so as to be outside of the influence of the atomic orbitals. During this process, we observed a drastic change in the field map with a gradual smoothing of the atomic scale structure. At a distance greater than 4 Å above the molecule, the field distribution becomes mainly sensitive to the global shape of the molecule.

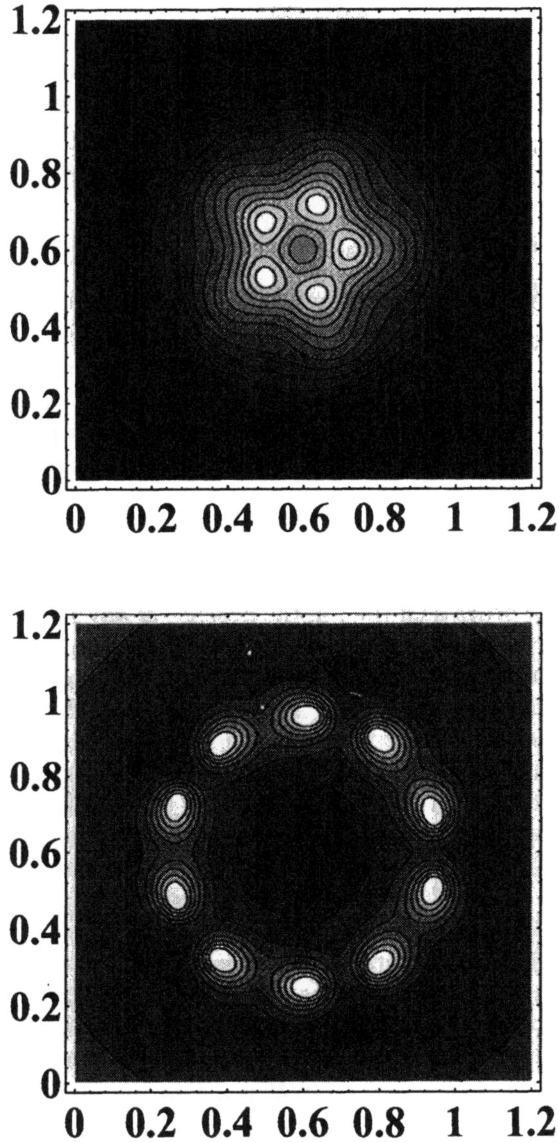

Fig.3: Grey scale representation of the modulus of the electric field $|E(X,Y)|$ calculated in the presence of a single C_{60} molecule adsorbed on a gold surface. The two axes are labelled in nm. The calculations are performed in two planes Z_0 = constant, namely: (upper field map) $Z_0 = 10.5$ Å; (lower field map) $Z_0 = 6.2$ Å. The tip–apex has been removed during the calculation. The applied field $\mathbf{E_0}$ is oriented along the OZ axis and the contour plots vary by steps of 0.015 V/Å.

4.2. Field maps generated by carbon capsules

Currently, carbon nanotubes can be fabricated from carbon vapor under ultrahigh vacuum conditions. These tubes consist of several concentric cylinders with different helicities. The smallest core of multishell tubes is considered to be a single–shell tube about 1 nm in diameter. Such tubes can be closed by hemispherical C_{60} caps. In this section we consider such tubular assembly of carbon atoms with the same radius as C_{60}. In fact, these elongated capsules, obtained by repeating the insertion of 10 or 18 atoms patterns between two C_{60} hemispheres can be built with different lengths. In the numerical work to be discussed in this section we just considered the case of a C_{260} molecule displaying an *armchair* structure(cf. Fig. 4).

Fig.4: Top view of the C_{260} molecule in the STM junction. This orientation has been chosen to perform the simulation described in Fig. 5.

To further explore the screening properties of such fullerene tubes lying in a biased STM junction, we have visualized in Fig. 5, a sequence of four field distributions. These four images have been calculated in the observation plane, defined by Z_0 = 6.5 Å, and passing through the center of the C_{260} molecule. We started the calculation (cf. map in the left upper corner of Fig. 5) with two plane metallic electrodes (without atomic apex), biased with an applied voltage V_0 = 2V and with a spacing D = 18 Å. The computational window is centered on the tubular part of the C_{260} molecule. The field observed outside the tubule drops over a short distance and appears to be strongly enhanced at the immediate proximity of the graphitic sheet. Furthermore our calculation indicates a strong and homegoneous depolarization phenomenon inside the tube over a large area. In this case, where the carbon cluster is embedded in the uniform applied field $\mathbf{E_0}$ perpendicular to the molecular axis, the electric field induced charge density reduces drastically the self–consistent field inside the carbon tubule (by about a factor five). We further tested the evolution of this screening effect when the C_{260} molecule is placed perpendicularly to the junction (i.e., when its main axis is parallel to the external field $\mathbf{E_0}$). In this configuration, in good agreement with recent first principle calculations [21], we observed a complete cancellation of the field in the tube like what we would obtain with a perfectly conducting tube of the same diameter.

In order to gain more insight into the influence of an highly inhomogeneous external field on this screening effect, we have introduced a (111) oriented metallic tip apex supported by the second electrode and located just above the center of the tubule. The three last images given in Fig. 5 show up the evolution of the field map upon reduction of the electrode spacing (see Fig. 5 caption). When the electrode supporting the metallic protrusion is brought near the tubular part of the fullerene, the previous electrostatic field distribution is locally modified. In spite of the screening phenomena described above, one still observes a great sensitivity of the structure of the field inside the carbon tubule with respect to the tip–sample spacing.

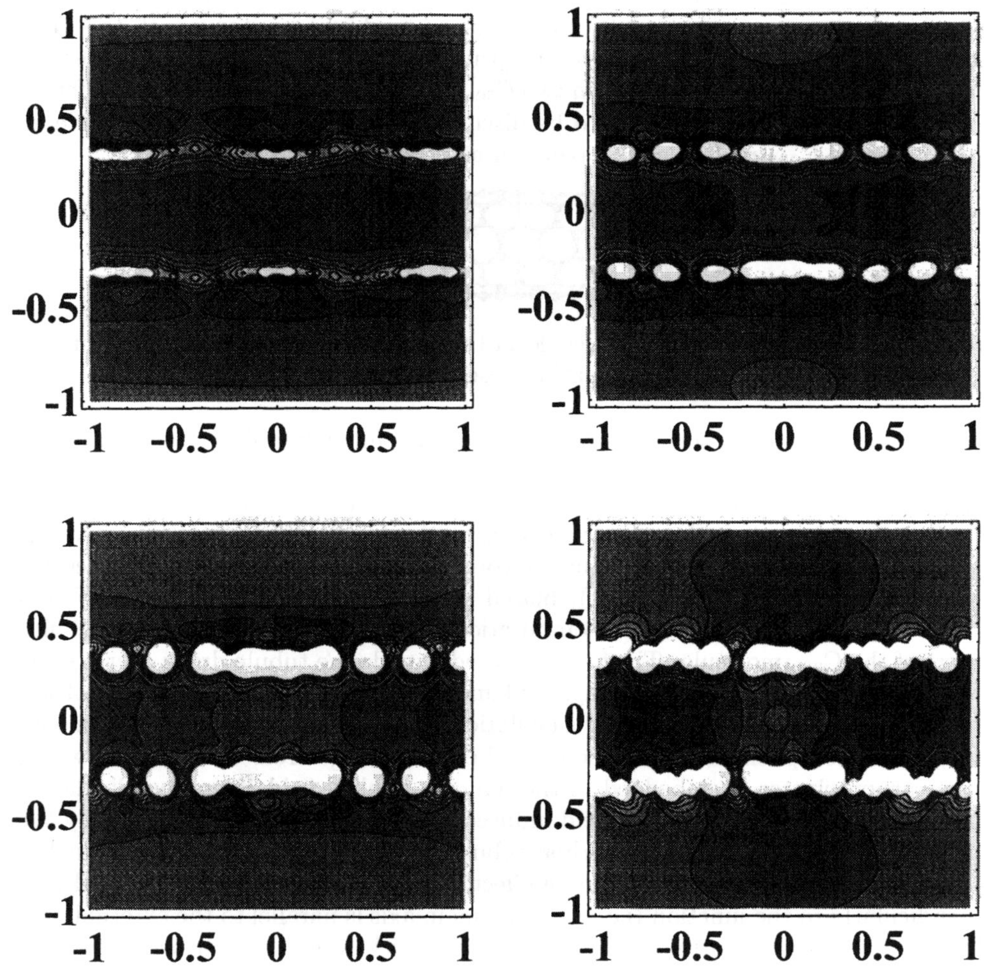

Fig.5: Sequence of grey contours level maps describing the evolution of the field inside the carbone tubule versus the tip–sample spacing. The two axes are labelled in nm.
(left upper corner map): calculation performed without microscopic tip apex and with a spacing $D = 18$ Å.
(right upper corner map): a microscopic tip is included in the calculation with a tip–sample spacing of 18 Å.
(left lower corner map): same with a tip–sample spacing of 16 Å.
(right lower corner map): same with a tip–sample spacing of 14 Å.

In Fig. 6, we visualize the field confinement effect observed inside the core of the tubule when the three–layers metallic apex is brought in its vicinity.

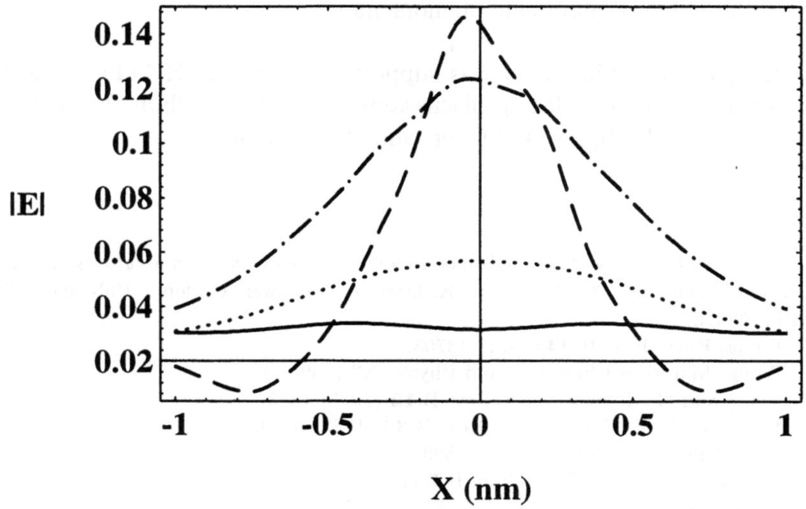

Fig.6: Scanlines describing the evolution of the field modulus (in V/Å) inside the carbon tubule along the main axis versus the tip–sample spacing.
(solid line): calculation performed without microscopic tip apex (D = 18 Å).
(dotted line): a (111) oriented microscopic tip is included in the calculation with a tip–sample spacing of 18 Å.
(dashed-dotted line): same with a tip–sample spacing of 16 Å.
(dashed line): same with a tip–sample spacing of 14 Å.

It may be seen on Fig. 6, that the field variation occurs on a scale fixed by the probe tip size itself. Typically for a tip–sample distance smaller than Z_0 = 14 Å, one observes a significant variation of the field intensity along the main axis, over a scale smaller than 5 Å. This effect is accompagnied with a field enhancement that varies between 1.5 and 6 when Z_0 decreases from 18 Å to 14 Å.

5. Conclusion

In the case of complex STM configurations, the numerical outputs generated by our EFP scheme show up that the field enhancement generated by the presence of molecules may be significantly increased thanks to a small sized metallic tip–apex. Although this effect appears to be very localized it can lead to complex depolarization effects around as well as inside long carbon tubules. The precise control of such phenomena could be a fundamental advantage when addressing the current problem of building nanoscale molecular switches triggered by the highly confined fields and currents controlable in STM devices.

Finally, with a relatively low cost in term of computer time, the numerical scheme derived from the EFP method appears to be an interesting tool to be coupled with other numerical codes (STM current, molecular adsorption energy, ...) and thus to facilate the interpretation and the optimization of recent experimental works devoted to the study of nanoscale phenomena.

Acknowledgments: This work was supported by the GDR 1145 of the CNRS. We have benefited of stimulating discussions with Ch. Joachim, Vu Thien Binh, O. J. F. Martin, L. Philippe, A. Mayer, and Ph. Lambin.

References

[1] *Nanosources and Manipulation of Atoms under high Fields and Temperatures: Applications*, edited by Vu Thien Binh, N. Garcia and K. Dransfeld (Kluwer Academic Publishers, Series E, Vol. **235** (1993)).
[2] T. T. Tsong, Phys. Rev. B **44** (1991) 13703.
[3] T. T. Tsong, Materials Chemistry and Physics **33** (1993) 1.
[4] T. T. Tsong and G. Kellogg, Phys. Rev. B **12** (1975) 1343.
[5] S. C. Wang and T. T. Tsong, Phys. Rev. B **26** (1982) 6470.
[6] N. D. Lang, Phys. Rev. B **45** (1992) 13599.
[7] N. D. Lang and W. Kohn, Phys. Rev. B **7** (1973) 3541.
[8] K. Hirose and M. Tsukada, J. Vac. Sci. Tech. B **12** (1994) 2164.
[9] K. Hirose and M. Tsukada, Phys. Rev. B **51** (1995) 5278.
[10] M. Devel, Ch. Girard, Ch. Joachim and O. J. F. Martin, Appl. Surf. Sci. **87/88** (1995) 390.
[11] P. A. Gravil, Ph. Lambin, G. Gensterblum, P. Senet, L. Henrard and A. A. Lucas, Surf. Sci. **165** (1995) 289.
[12] M. Devel, Ch. Girard and Ch. Joachim, Phys. Rev. B **53** (1996) (pages to be provided).
[13] J. Mahanty and B. W. Ninham, in *"Dispersion Forces"* (Academic Press, London, 1976).
[14] P. E. Blöchl, J. Chem. Phys. **103** (1995) 7422.
[15] Ch. Girard, A. Dereux and O. J. F. Martin, Surf. Sci. **295** (1993) 445.
[16] E. I. Altman and R. J. Colton, Surf. Sci. **279** (1992) 49.
[17] R. Berndt, R. Gaisch, J. K. Gimzewski, B. Reihl, R. R. Schlittler, W. D. Schneider and M. Tschudy, Science **262** (1993) 1425.
[18] S. Behler, H. P. Lang, S. H. Pan, V. Thommen-Geiser, H.-J. Güntherodt, Z. Phys. B **91** (1993) 1.
[19] T. Sakurai, X. D. Wang, T. Hashizume, V. Yurov, H. Shinohara and H. W. Pickering, Appl. Surf. Sci. **87/88** (1994) 405.
[20] M. Froimowitz, J. Comput. Chem. **12** (1991) 1129.
[21] L. Lou, P. Nordlander and R. E. Smalley, Phys. Rev. B **52** (1995) 1429.

A PROPOSAL OF ATOM ELECTRONICS BASED ON ATOM/MOLECULE SWITCHING DEVICES

Y. WADA
Advanced Research Laboratory, Hitachi, Ltd.
2520 Akanuma Hatoyama, Saitama 350-03 Japan

This paper proposes the candidate devices to supersede MOSFET (Metal-Oxide-Semiconductor Field Effect Transistor) for the future information processing, Atom Relay Transistors (ART) and Molecular Single Electron Switching Transistors (MOSES]. These atom/molecule devices are with total dimensions well below a few nm and an operation speed of more than Tera (10^{12}) Hz. Their performances are evaluated on the basis of the characteristics necessary for integrated circuit devices, and are found to be the most promising candidates among the nanoscale devices including quantum devices and superconductor devices. The basic technology development toward the realization of ART and MOSES are reported, such as Beam Assisted Scanning Tunneling Microscope (BASTM) for insulator observation, Needle Formation and Tip Imaging (NFTI) for atom-scale evaluation of STM tip apex and Micromachine STM (μ-STM) for direct evaluation of the vacuum tunneling gap between tip and sample, together with the atom wire fabrication technology development.

1. **Introduction**

Advances in silicon integrated circuit technology have been made possible by the down scaling of metal oxide semiconductor (MOS) transistors for higher performances [1]. However, several physical and chemical limitations might prove obstacles to micro-miniaturization beyond the 0.1 μm technology level [2]. Innovative MOS transistor structures, it is claimed, makes further down scaling possible [3], though improvement would be by a factor of two at most. The trend curves of MOS transistor down scaling and integration of large scale integrated circuit (LSIs) are shown in Fig. 1, in which device integration has increased by a factor of four every three years, while the minimum dimension has been reduced by a factor of 0.7 every three years. They also indicate that the minimum dimension of about 0.1 μm might be reached around the year of 2010. Therefore nano scale devices, such as quantum effect devices [4], have recently been attracting attentions as candidates to replace silicon MOS transistors for future information processing integrated circuit devices. However, it is uncertain whether those "quantum" devices will be able to replace MOS transistors, because the characteristics of these devices are not necessarily be superior to those of MOS transistors, which are almost ideal as an information processing device, as will be shown later.

This paper proposes atom size switching devices, Atom Relay Transistor (ART) [5] and Molecular Single Electron Switching Transistor (MOSES) [6], with dimensions

C. Joachim and S. Roth (eds.), Atomic and Molecular Wires, 193-202.

well below a few nm, and an operation speed of more than 10^{12} (Tera:T) Hz. Operation principles, simulated device characteristics and performance evaluation are shown as well as the technology development towards the realization of these devices.

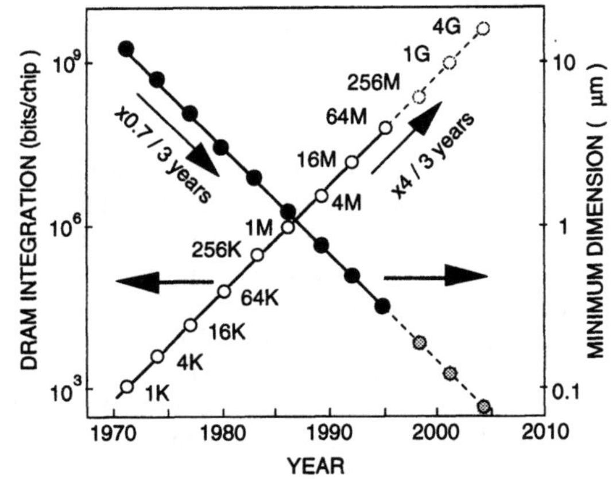

Figure 1. Trend curves of MOS transistor down scaling and DRAM LSI integration

2. Characteristics necessary for information processing devices

Generally, integrated circuit devices should fulfill the following requirements for high performance information processing [5]. They are (a) input and output (I/O) signal balance, (b) I/O signal isolation, (c) high switching speed, (d) dense integration capability and (e) fabricability. Other factors, such as operation margin, noise immunity, power dissipation and reliability have to be taken into consideration in the integrated circuit design. Among these five factors, (c) high switching speed has usually been the only criterion for evaluating newly proposed devices, however, it is essential to fulfill these five factors in a balanced manner. The reason why MOS transistors have been used in integrated circuits so far and are expected to be used down to their limit of around 0.1 μm, is that they almost ideally fulfill the above five requirements. These requirements will be the basis of any future information processing devices, even if computer architecture is changed. For example, massively parallel computers will operate faster, if the individual processor operates faster.

3. Atom/molecule switching devices - prospect for the future

This section explains the operation principles of ART and MOSES and considers whether the proposed devices can fulfill the above mentioned five requirements.

3.1. ATOM RELAY TRANSISTOR (ART)

3.1.1. *Principle of Operation*

The basic concept of ART [5] is schematically depicted in Fig. 2, in which an individual atom is represented by a circle. In the figure, ART is depicted as consisting of an atom wire, a switching atom, a switching gate and a reset gate. The operation principle is as follows. The atom wire is conductive (on) when the switching atom is set in the atom wire, and it is non-conductive (off) when the switching atom is removed from the atom wire [5]. The movement of the switching atom is made possible by the electric field supplied from the switching gate, and the first requirement of (a) I/O balance for ART will be fulfilled. The on/off characteristics are simulated by the tight binding method and the results are depicted in Fig. 3, in which 250 atoms are arranged in a straight line with a separation of 0.2 nm. One electron is injected from the far left side of the atom wire with an electric field of 3 x 10^5 V/cm, and the electron density flowing in the atom wire is shown with a time interval of 6 fs. The results clearly indicate that ART exhibits on/off switching characteristics depending on the position of the switching atom. The second requirement, (b) I/O signal isolation, is also fulfilled, because ART is a three terminal device.

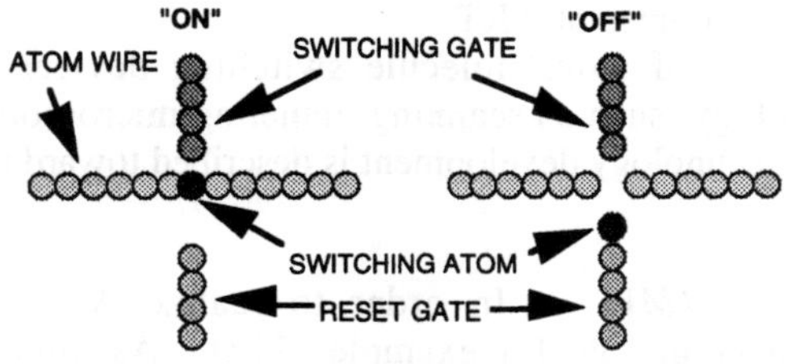

Figure 2. Basic configuration of Atom Relay Transistor (ART) [5]

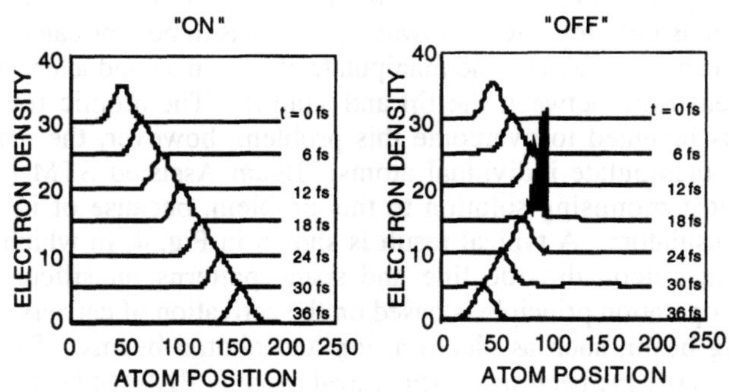

Figure 3. Simulated "ON" "OFF" switching characteristics of ART [5]

3.1.2. *Switching Speed*

The switching speed of the switching atom from the "on" site to the "off" site, which corresponds to the switching frequency of ART, was simulated at around 30 THz by the first principle method [7]. The transmission velocity of the electric field also allows operation of ART at above 10 THz. Therefore the overall performance of ART should well exceed 10 THz. Thus the third condition, (c) very high switching speed, has been fulfilled.

3.1.3. *Memory Cell and Logic Circuits*

The basic configuration of ART shown in Fig. 2 suggests that ART can constitute a static memory cell by itself. In addition, ART offers dynamic operation by using the self-relay configuration, in which the switching atom is removed from the atom wire by the electric field from the atom wire itself and electrons are trapped in the loop [8]. A dynamic memory cell uses only about 10^3 times the atom area, and a 10^9 bit occupies only about 200 μm square of area. The trapped electron might disappear through leakage passes such as insulator leakage, tunneling through the gap or radiation of light. A lifetime evaluation of the trapped electron would constitute a very interesting subject for scientific study. NAND and NOR gates are also designed [5], and von Neumann type computer can be constructed. Rule of thumb estimation of the area occupied by a 10^7 gate logic circuit and a 10^9 bit memory, which corresponds to almost the same integration as the state-of-the-art supercomputers, results in a 20 μm square area and 200 μm square area, respectively. Therefore the total area occupied by the ART supercomputer would be about 200 μm square, and even the longest signal delay on the chip would be well below the order of 10^{-12} s, which is short enough even for operation at more than 1 THz. Thus, the fourth condition, (d) dense integration, would be fulfilled.

3.1.4. Fabrication Technologies of ART

Fabrication technologies of atom/molecule switching devices would be built on scanning probe technology, such as scanning tunneling microscope (STM) [9]. Here, the current fabrication technology development is described toward the materialization of atom electronics.

Beam Assisted STM (BASTM). In order to realize ART, it is necessary to manipulate individual atoms by, for example, STM. As already indicated in the literature, STM can manipulate the individual atom on a conducting solid surface [10]. Therefore by using this technology, it should be possible to realize ART, and the logic and memory circuits utilizing ART. However, ART has to be fabricated on an insulator substrate. STM can only observe and manipulate atoms on a conductor surface, because it requires current flow between the tip and sample. The atomic force microscope (AFM) [11] was invented to overcome this problem, however, the resolution is not high enough to manipulate individual atoms. Beam Assisted STM (BASTM) [12] would be the most promising solution to this problem, because of its capability for observation of insulators. A typical result is shown in Fig. 4, in which 100 nm wide and 10 nm thick silicon dioxide line and space patterns on silicon substrate are delineated. The operation principle is based on the activation of carriers in the insulator by an irradiating beam, such as electron, ion and photon beams. Thermally grown silicon dioxide layers have been successfully evaluated by BASTM [13], which confirms the possibility of manipulating atoms on insulators.

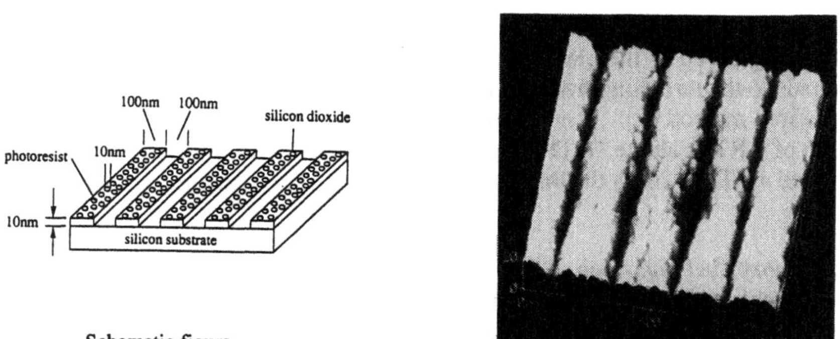

Schematic figure

Figure 4. BASTM observation of a 100 nm wide and 10 nm thick silicon dioxide layers

Needle Formation and Tip Imaging (NFTI). In order to manipulate individual atoms by STM, it is necessary to use tips with a very controlled apex. The Needle Formation and Tip Imaging (NFTI) method [14,15] delineates the three-dimensional tip apex image and in-situ evaluation of the apex is made possible. The nano needle structure formed by NFTI is shown in Fig. 5, in which the diameter and height measures 1.5 and 9 nm, respectively. Using the nano needle as the probing tip, the STM tip apex structure was delineated, and the typical result is shown in Fig. 6. The atomic protrusion on the tip apex is clearly indicated, which made it possible to detect the surface structures at an atomic resolution [8]. This is the first direct experimental evidence of the theoretical and experimental hypothesis that an atomic protuberance on the tip apex is indispensable to obtain an atomic resolution.

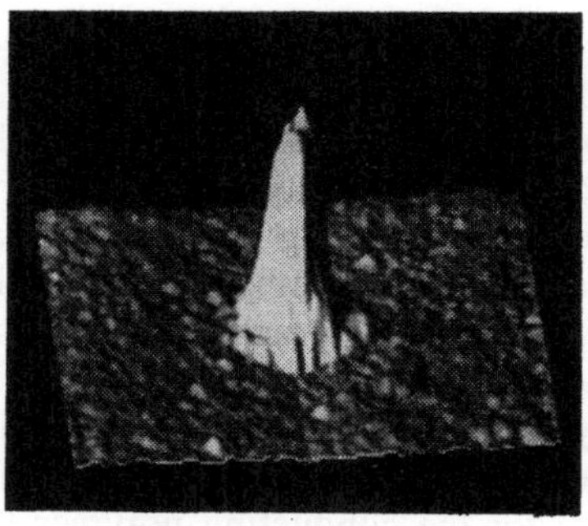

Figure 5. **Nano-needle structure formed by NFTI**

Figure 6. **STM tip apex observed by NFTI**

Atom Wire Fabrication. The first step to demonstrate the operation of ART would be to measure the conductance of an atom wire. An example of the fabricated atom wire is shown in Fig. 7, in which hydrogen passivated silicon (100) surface was used as a starting material (a). One hydrogen row was removed by an STM modification method to construct dangling bond wire (b) [16]. By applying metal atoms on the dangling bond structure, the dangling bond wire was filled with metal atoms to ascertain the electrical conduction (c). Thus the conductance of a single atom wire would be measured if contact pads are formed on both ends of the wire [17].

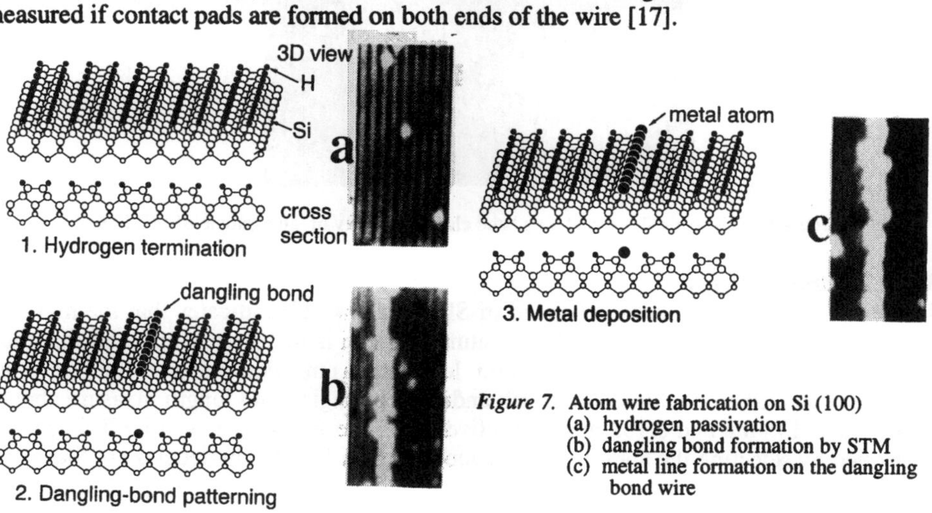

Figure 7. **Atom wire fabrication on Si (100)**
(a) hydrogen passivation
(b) dangling bond formation by STM
(c) metal line formation on the dangling bond wire

3.2. MOLECULAR SINGLE ELECTRON SWITCHING (MOSES) DEVICE

Molecular devices attracted much attention when they were proposed in 1980 by Carter [18], however, the interest gradually faded away due to several causes such as inaccessibility to individual molecules and the slow switching speed of molecules for an information processing device. Recent progress in single electron tunneling (SET) devices [19] attracts attention because of its potentially superior operation characteristics. The SET devices operate faster, if the quantum dot size is reduced to 1 nm [20]. However fabrication as well as size control should become almost impossible by a conventional fabrication technology in a nm regime. Molecular systems should realize much better dimensional control, because the dimension of a molecule is determined by the number of bonds within the molecule.

3.2.1. *Principles of Operation*

A MOSES device consists of conducting polymer and insulating polymer, as schematically shown in Fig. 8 [6]. In the figure, polyacetylene and polyethylene molecules were chosen as typical conducting and insulating molecules, respectively. The insulator part acts as a tunnel barrier, and the conducting polymer island as a quantum dot. The electron density of the polyacetylene-polyethylene-polyacetylene molecular structure was calculated by a molecular orbital (MO) method, and the results are also demonstrated in Fig. 8. These results clearly indicate that the conduction electron density is negligible in the insulator (polyethylene) part while it is high in the conductor (polyacetylene) part of the molecule. Recently, a relatively solid and straight chain molecule has been proposed as a candidate for the conducting polymer [21]. Therefore, these molecules might be the candidates for realizing MOSES. In addition, three terminal MOSES devices should also be realizable by synthesizing an appropriate molecular structures [22] with branches, dots and gates.

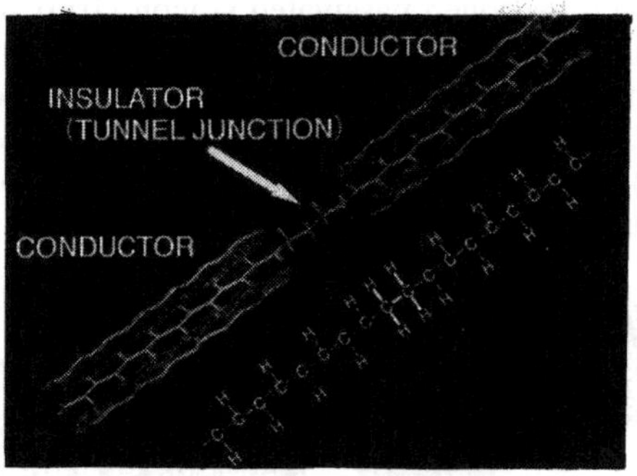

Figure 8. Schematic figure and simulated electron density of an example of MOSES

3.2.2. *Operation Speed*

The simulated ultimate operation speed of SET devices [20] indicates that the highest speed is in the order of 1 THz if the quantum dot size is less than 1 nm, and that the tunnel barrier width must also be very thin, less than 1 nm. For example, the size of a benzene ring, carbon double bond length and carbon single bond length measure about 0.242 nm, 0.135 nm and 0.153 nm, respectively. Therefore, the size of MOSES can be controlled by selecting the appropriate molecular design and that molecules are likely to be the most appropriate material for realizing high performance SET devices. The ultimate speed of a MOSES device should be of the order of 10 THz, as indicated in ref. [20], if the quantum dot size is reduced to about 0.3 nm, such as benzene ring.

3.3.3. Memory and Logic Circuits

Memory and logic circuits can be formed by using MOSES devices, in which conventional SET device and circuit design [23] can be applied. Therefore, no new circuit design should be necessary. Appropriate molecular design should achieve the necessary circuit characteristics, including multiple output wiring.

3.3.4. Fabrication Technologies of MOSES Devices

The first major difficulty towards the realization of MOSES would be the access to individual molecules, as mentioned previously in this section. Recently, a novel scanning tunneling microscope (STM) manipulation method was proposed using micromachine fabrication technology (μ-STM) [24]. The SEM micrograph of the fabricated μ-STM is shown in Fig. 9. The vacuum tunneling gap between the μ-STM tip apex and the sample was successfully observed by a transmission electron microscope (TEM), as shown in Fig. 10 [25]. The two electrodes are made with gold, and thiol groups on both ends of the conducting molecule should easily make contacts to the gold electrodes. Therefore the electrical conduction characteristics of the molecule should be reliably measurable, and the high performance operation of MOSES should be demonstrable.

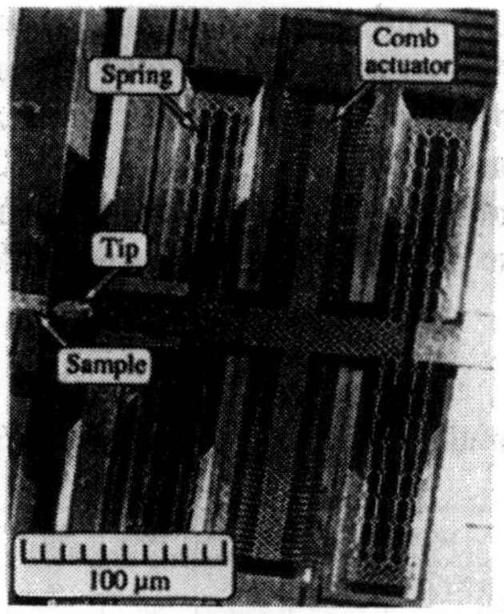

Figure 9. SEM micrograph of the fabricated μ-STM

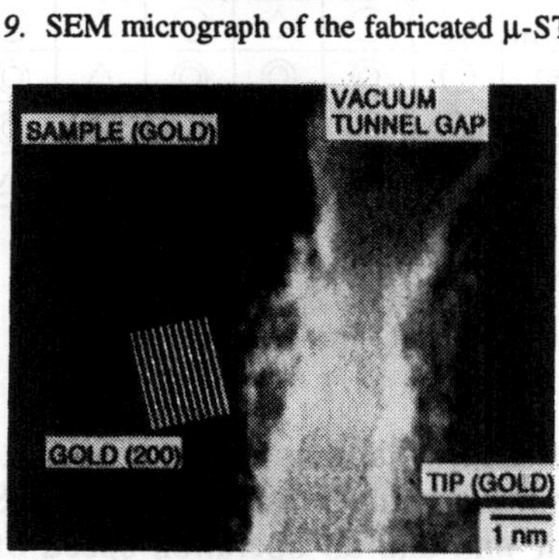

Figure 10. TEM micrograph of the vacuum tunneling gap

4. Evaluation of Candidate Devices for the Future

The candidate switching devices for the next decade, including quantum devices, conventional single electron devices, conventional molecular devices, ART and MOSES devices, were evaluated according to the five indispensable factors listed in section 2, and the results are summarized in Table 1. It is indicated that MOS transistors almost ideally fulfill all the requirements, and that ART and MOSES are the most promising candidates to supersede them. The candidate nano scale devices listed in Table 1, in addition to the historical information processing electron devices, were evaluated semi-quantitatively in terms of speed and integration capability, and the results are summarized in Fig. 11. Human beings started electronic information processing using mechanical relay. The very slow and unreliable operation of mechanical relay (mechanical electronics) was soon replaced by a vacuum tube [25] (vacuum electronics). Then the invention of transistors [26] revolutionized the computer, and solid state electronics devices including integrated circuits [27] dominated electronic information processing. As indicated in Fig. 11, the innovation, or paradigm shift (e.g. from mechanical electronics to vacuum electronics to solid state electronics), has been accomplished because the performances of the latter devices are superior to those of preceding ones by at least two orders of magnitude.

Now that solid state electronics devices are approaching their limits, candidate devices are expected to supersede them. Conventional molecular devices exhibit very high integration density, however, their operating speed is extremely slow and they cannot be prime candidates for future devices. Quantum devices demonstrate better characteristics in terms of integration and switching speed, however, they do not exhibit orders of magnitude higher switching speed or integration than the present silicon MOS transistors. Single electron devices would have to be scaled down to less than 1 nm in order to obtain very high performances. Therefore ART and MOSES are likely to be the most promising candidate devices for future electronic information processing integrated circuits, because of their potentially very fast operation speed and dense integration characteristics.

Table 1. Comparison of the performances of candidate information processing devices

	MOS	QUANTUM DEVICE	SUPER CONDUCTOR DEVICE	SET	MOLECULAR DEVICE	ART	MOSES
I/O BALANCE	○	△	○	○	△	○	○
I/O ISOLATION	○	○	△	○	△	○	○
SPEED	○	◎	◎	◎	△	◎	◎
INTEG-RATION	○	○	○	○	◎	◎	◎
FABRI-CABILITY	○	△	△	△	△	△	△

5. Conclusions

This paper proposed nano scale switching devices, Atom Relay Transistors (ART) and Molecular Single Electron Switching Transistors (MOSES), for the next score, with dimensions of only a few nm, and an operation speed of more than 10 THz. The switching characteristics of ART were demonstrated by simulation. Fundamental logic circuits such as NAND and NOR gates and memory circuits were proposed, which can integrate a supercomputer in an area 200 μm square, with 10^7 gates of logic circuit and 10^9 bits of memory, and operate at more than 1 THz. MOSES device structures were also demonstrated by simulation and very dense and fast operation characteristics were foreseen. New technologies indispensable for the fabrication of these nano scale devices were demonstrated, including BASTM, NFTI and μ-STM. Experimental results to fabricate atom wire structures were also demonstrated. Various candidate nano scale devices were evaluated on the basis of the characteristics necessary for future integrated circuit devices, and ART and MOSES devices were found to be the most promising devices for future information processing integrated circuits.

The author would like to express his sincere gratitude to Dr. Shojiro Asai for his support and continuous encouragement throughout the work. Dr. Tsuyoshi Uda, Dr. Satoshi Watanabe and Dr. Takashi Goda are acknowledged for fruitful discussions on the structure and device performance simulation. He would also like to express his thanks to Dr. Munehisa Mitsuya, Dr. Tomihiro Hashizume, Dr. Mark I. Lutwyche, Mr. Masayoshi Ishibashi, Ms. Nami Sugita, Mr. Seiji Heike, Mr. Ken Nakajima and Mr. Taro Hitosugi for their contributions to the work. ULSI Research Department and Electron Device Department of Central Research Laboratory are acknowledged for fabricating the chips.

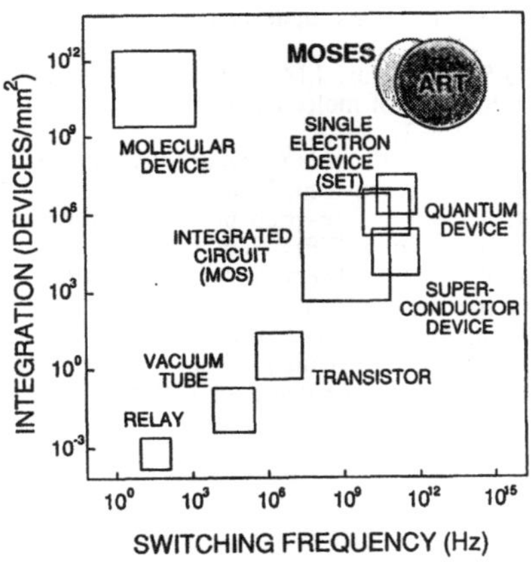

Figure 11. Comparison of the switching speed and integration density of historical and candidate information processing devices

6. References

[1] Dennard, R.H., Gaensslen, F.H., Yu, H., Rideout, V.L., Bassous, E. and LeBlanc, A.L. (1974) Design of ion implanted MOSFETs with very small physical dimensions, *IEEE J. Solid St. Circuits* **SC-9**, 256.

[2] Sugano, T. (1990) *VLSI Process Data Handbook*, Science Forum, Tokyo
Keyes, R.W. (1992) The future of the transistor, *Physics Today*, **45**, 42 .

[3] Fiegna, C., Iwai, H., Wada, T., Saito, M., Sangiorgi, E. and Ricco, B. (1994) Scaling the MOS transistor below 0.1 mm: methodology, device structures, and technology requirements, *IEEE Trans. Electron Devices*, **41**, 941.

[4] B. L. Al'tshuler, P. A. Lee and R. A. Webb (Eds.) (1991) *Mesoscopic Phenomena in Solids*, North-Holland, Amsterdam.

[5] Wada, Y., Uda, T., Lutwyche, M., Kondo, S. and Heike, S. (1993) A proposal of nanoscale devices based on atom/molecule switching, *J. Appl. Phys.*, **74**, 7321.
[6] Wada, Y. (1994) A proposal of molecular single electron devices, *Trans. IEICE*, **OME 93-54**, 31.
[7] von Allmen, P. and Hess, K. (1995) Dynamics of an atomic switch computed by first-principles molecular dynamics, *Phys. Rev. B* **52**, 5243
[8] Wada, Y. (1996) Atom electronics:A proposal of nano-scale devices based on atom/molecule switching, *Microelectronic Engineering*, **30**,375
[9] Binnig, G. and H. Rohrer, Gerber, Ch., and Weibel, E (1982) Surface studies by Scanning Tunneling Microscopy, *Phys. Rev. Lett.*, **49**, 57.
[10] Eigler, D.M., and Schweizer, E.K. (1990) Positioning single atoms with a scanning tunneling microscope, *Nature*, **344**, 524.
[11] Binnig, G., Quate, C.F. and Gerber, C. (1986) Atomic force microscope, *Phys. Rev. Lett.*, **56**, 930.
[12] Heike, S., Wada, Y., Kondo, S., Lutwyche, M.I., Murayama, K. and Kuroda, T. (1994) Scanning tunneling microscope measurement of insulator surfaces, *Appl. Phys. Lett.*, **64**, 1100.
[13] Heike, S., Wada, Y., Kondo, S., Lutwyche, M.I., Murayama, K. and Kuroda, T. (1995) Evaluation of thin silicon dioxide layers by beam assisted scanning tunneling microscope, *Japan. J. Appl. Phys.*, **34**, 110 .
[14] Heike, S., Hashizume, T. and Wada, Y. (1995) In situ direct imaging of scanning tunneling microscope tip apex, *Japan. J. Appl. Phys.*, **34**, L1061.
[15] Heike, S., Hashizume, T. and Wada, Y. (1996) Nano-Needle Formation on Silicon (111) Surface using a Scanning Tunneling Microscope Tip, *J. Appl. Phys.*, in press.
[16] Watanabe, S., Ono, Y., Hashizume, T. and Wada, Y. (1995) Electronic Structure of an Atomic Wire on a Hydrogen-Terminated Si (111) Surface : First Principle Study, *Phys. Rev. B*, **52**, 10768.
[17] Hashizume, T., Heike, S., Lutwyche, M., Watanabe S. and Wada, Y. (1996) Development of an in-situ Method of Fabricating Artificial Atom Structures on Si (100) Surfaces and their Connection to Bulk Electrode, *Proc. International Conference on Solid State Devices and Materials*, Japan Society Appl. Phys., Yokohama.
[18] Hong, F.T. (Ed.) (1989) *Molecular Electronics*, Plenum Press, New York.
[19] Likharev, K.K. (1988) Correlated discrete transfer of single electrons in ultrasmall tunnel junctions, *IBM J. Res. Develop.*, **32**, 144.
[20] Lutwyche, M.I. and Wada, Y. (1994) Estimate of the ultimate performance of the single-electron transistor, *J. Appl. Phys.*, **75**, 3654.
[21] Pearson, D.L., Schumm, J.S. and Tour, J.M. (1994) Iterative divergent/convergent approach to conjugated oligomers by a doubling of molecular length at each iteration. A rapid route to potential molecular wires, *Macromolecules*, **27**, 2348.
[22] Ulman, A (1991) *An Introduction to Ultrathin Organic Films*, Academic Press, San Diego.
[23] Yano, K., Ishii, T., Hashimoto, T., Kobayashi, T., Murai F. and Seki, K. (1993) A room temperature single-electron memory device using fine-grain polycrystalline silicon, *Technical Digest, IEEE International Electron Devices Meeting*, IEEE, New York p. 541.
[24] Lutwyche, M. and Wada, Y. (1995) Observation of a vacuum tunneling gap in a transmission electron microscope using a micromechanical tunneling microscope, *Appl. Phys. Lett.*, **66**, 2807.
[25] Spangenberg, K.R. (1948) *Vacuum Tubes*, McGraw-Hill, New York.
[26] Bridgers, H.E., Scaff, J.H. and Shive, J.N. (Eds.), (1958) *Transistor Technology*, Nostrad, Prinston.
[27] Kilby, J., (1976) The invention of the integrated circuit, *IEEE Trans. Electron Devices*, **ED-23**, 648.

MOLECULAR WIRES: RESONANCES, STAIRCASES, RECTIFICATION, BONDING AND SPECULATION

MATHIEU KEMP*, VLADIMIRO MUJICA**, ADRIAN ROITBERG***, YI MAO AND MARK A. RATNER
Department of Chemistry and Materials Research Center
Northwestern University
Evanston, Illinois 60208, USA

ABSTRACT:

Electronic structure models for computation of conductances in molecular wire circuits, as well as some actual calculational results and comparison with experiment, are presented. The conductance in a molecular wire circuit is related to the rate of long range nonadiabatic electron transfer, but important differences arise from the coupling to the electronic structure continuum in the conductance compared to the vibronic quasi-continuum in the long range electron transfer problem. Scattering formulas based on the Landauer approach permit calculation of resonance, rectification, and electronic level effects. Fluctuation phenomena can also be included within a simple extended Huckel model for the molecular electronic structure, but coulomb staircases and single electron tunneling behavior require introduction of repulsions on the molecular wire itself. Some experimental comparisons are given, and speculations with respect to gates, transistors, extended wires and vibronic mixings are ventured.

1. Introduction

As some of the beautiful experimental contributions in this book make clear, the measurement of currents through atomic and molecular wires is rapidly becoming experimentally approachable, and a number of pioneering measurements have now been reported.[1-4] For molecular wires, the current can only be understood in terms of a composite structure involving leads (electrodes) and the molecular structure itself. Measurements using both scanning tunneling microscopies and mechanical break junctions are becoming available,[6] and their understanding in terms of molecular electronic structure concepts promises both to expand our understanding of charge transfer and transport in molecules, and (perhaps) to illuminate some interesting applications in molecular electronics and control of charge transport.[7]

Permanent Address: *Departamento de Física, Universidad de los Andes, A.A. 4976, Bogota, Colombia; **Universidad Central de Venezuela, Facultad de Ciencias Escuela de Química, Apartado 47102, Caracas 1020A, Venezuela; ***National Institute of Standards and Technology, Biotechnology Division, Building 222, A-353, Gaithersburg, MD 20899.

C. Joachim and S. Roth (eds.), Atomic and Molecular Wires, 203-217.

Conduction in wires is related to long range non-adiabatic intramolecular electron transfer. This is a problem of substantial interest in chemistry and a great deal of progress has been made in understanding it.[7-12] In these reactions, charge is transfered within the molecule and the rate can be well understood using a generalization of the Golden Rule, in which the electronic states are coupled to vibronic continuum (comprising both reorganization of the solvent environment and reorganization of the intramolecular coordinates). Thus, the rate constant can be written as

$$k_{ET} = \frac{2\pi}{\hbar} V_{RP}^2 (DWFC) \tag{1}$$

Here k_{ET} is the first order rate constant (units of inverse time) for the transfer, V_{RP} is the nonadiabatic electron transfer matrix element that links the electronic state of the reactant and that of the product geometry, while DWFC denotes a density of states weighted Franck-Condon factor, that characterizes the vibrational reorganization and the energy transfer between the electronic and vibrational states.

In a molecular wire circuit charge moves between the two metallic electrodes through the electronic structure of a molecule. The striking contrast between the conductance and k_{ET} really centers on three phenomena: in the conductance problem the electrodes define very clearly the initial and final states, the applied voltage W provides a continuous energy difference variation between reactants and product, the continuum that couples to the discrete electronic states is the electron continuum of the bands in the electrode (rather than the vibronic continuum seen for nonadiabatic electron transfer).

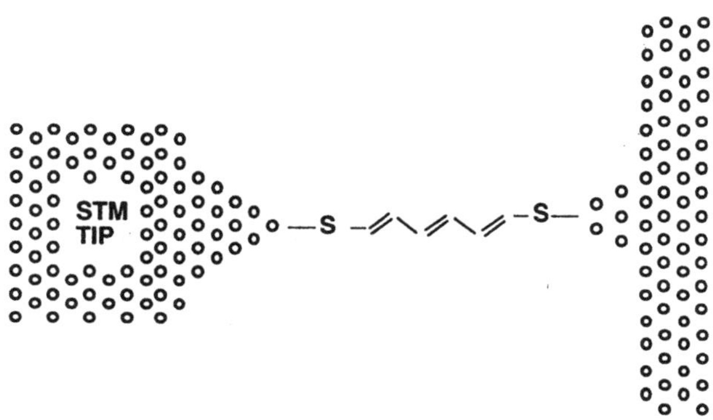

Fig. 1. Schematic representation of a molecular wire. The device consists of a conjugated polymer sandwiched between two metallic electrodes. For concreteness, we display an example where the top electrode is an STM tip, and the wire is adsorbed on the bottom electrode. When the STM tip is in close contact with the top of the wire, current flows between the electrodes through the wire.

For the characteristic molecular wire circuit sketched in Fig. 1, application of scattering method to compute the conductance has been given by a number of workers.[5,13-29] In the small applied voltage regime, where ohmic behavior is observed, the conductance is given by

$$g = \frac{j}{W} = \frac{2e^2}{\pi\hbar} |G_{1N}|^2 \Delta_1(E_F)\Delta_N(E_F) \tag{2}$$

Here g, j, W are respectively the conductance, the current and the applied potential. G_{1N} is the element of the molecular electronic Green's function between the first and Nth (last) sites of the molecule. The spectral densities Δ_1 and Δ_N are properties of the reservoirs and depend parametrically on the chemisorption energies. This equation can be derived in several ways, notably from the Bardeen tunneling formula,[30] or from a simple scattering approach[5,13-29]. The Newns-Anderson model[31] is used to describe the electrodes as simple, one dimensional continua connected to the molecule by one point (as shown in Fig. 1). The details of the formula are given in Section II below.

When the applied voltage is large, ohmic behavior may not hold; under these conditions, the current can be directly defined as

$$j = \frac{2e}{\pi\hbar} \int_{E_F - eW}^{E_F} dE \Delta_1(E)\Delta_N(E + eW)|G_{1N}(E,W)|^2 \tag{3}$$

Here the integral runs over the energy interval between the Fermi energy on the acceptor and the Fermi energy minus the potential difference on the donor.

These preliminary theories ignore a number of striking phenomena (vibronic trapping, photoexcitation, orientational gating and control) to obtain a useful calculational technique for molecular wires. This has been done by a number of workers.[13-29]

We can relate the simple Tersoff-Hamann[32] limit for the intensity of STM imaging to the conductance formulas that we have written. The Tersoff-Hamann result simply says that the current is proportional to $\rho\ (E_F,r)$ - that is, the density of states at the Fermi energy at the position of the acceptor tip. When there is a kinetic limitation that intervenes between donor and acceptor, such as occurs with coulomb staircases[33] or the molecular wire features, the Tersoff-Hamann results should fail; thus molecular wire behavior relates to a large number of interesting chemical and physical phenomena (the long range electron transfer, STM imaging, surface electro-chemistry).

In this overview, we will indicate a few of the observations that have been obtained from our own work, based on electronic structure theory and the conductance formulas of Eqs. 2 and 3. Section 2 provides a quick overview of the formalism to be used. In Section 3, we examine the simplest observable phenomena, including resonant tunneling through molecular eigenstates, length and geometry dependence of the conductance, the nature of the spectrum for conductance as a function of applied voltage or Fermi energy (the relationship between the eigenvalue structure of the isolated molecule and the transport). We also distinguish superexchange type transfer through electron and hole levels within the molecular wire itself.

In Section 4, we examine the effects of diagonal disorder (analogous to that that causes localization in the Anderson picture[34] of disordered systems). We will study the effect of disorder of molecular site energies on the current. The important observations are that the conductance obeys a log normal statistical distribution, and localization can indeed occur due to disorder in the energies. Some non-intuitive effects, including disorder induced conductance, are observed.

In Section 5, the effects of inter-electronic coulomb repulsion are included. These are responsible for variations in the population of electrons in the molecular wire as a function of applied voltage. Coulomb and eigenvalue staircases are observed, and large steps in the conductance are seen as a function of applied voltage. Rectification behavior can also be easily observed.

In Section 6, we comment on the experimental comparison. It is striking that the measured resistance[1-4] in molecular wire circuits, generally of the order 10 MΩ, is quite close to those calculated using a fairly simple extended Huckel model[35] for the molecular electronic structure coupled with the scattering theory forms of Eqs. 2 and 3. This is in striking contrast with atomic wires, where the conductance is at least 10^3 higher, and quantization steps are often seen.[6,36]

Finally, Section 7 deals with speculations on experiments that have not yet been reported, and theory that has not yet been completed.

2. Formalism: Conductance and Current Formulas Based on the Scattering Theory Approach

The formalism used is based on the Newns-Anderson model for chemisorption and an extended Huckel model[35] for the molecular wire itself. The conductance is given by Eqs. 2 and 3 for respectively, small and large applied potentials, W. The Green's function matrix is determined by

$$(E-H)^{-1} = G(E) \tag{4}$$

Here H is the matrix corresponding to the electronic structure Hamiltonian, which is modified only in the first and last elements (that is, in the elements adjacent to the terminal site on the electrodes). These readjustments are given by

$$H_{11} = H_{11}^{(H)} - \Sigma_1 (E) \tag{5a}$$

$$H_{NN} = H_{NN}^{(H)} - \Sigma_N (E) \tag{5b}$$

Here the $^{(H)}$ denotes the extended Huckel Hamiltonian for the molecule itself, while the self energies have real and imaginary parts, and are fixed by

$$\Sigma_K (E) = \Lambda_K(E) - i\Delta_K(E) \qquad K = 1,N \tag{6}$$

Where

$$\Lambda_K(E) = \frac{P}{\pi}\int \Delta_K(E')\frac{dE'}{E-E'} \tag{7}$$

Where P is the principal part, and

$$\Delta_K(E) = \begin{cases} \dfrac{\beta_K^2}{\gamma}\dfrac{E}{2\gamma} & \left|\dfrac{E}{2\gamma}\right| < 1 \\[2ex] \dfrac{\beta_K^2}{\gamma}\left\{\dfrac{E}{2\gamma} + \sqrt{\dfrac{E^2}{4\gamma^2} - 1}\right\} & \dfrac{E}{2\gamma} < -1 \\[2ex] \dfrac{\beta_K^2}{\gamma}\left\{\dfrac{E}{2\gamma} - \sqrt{\dfrac{E^2}{4\gamma^2} - 1}\right\} & \dfrac{E}{2\gamma} > 1 \end{cases} \tag{8}$$

These results are taken from the analysis by Newns.[31] Note that choice of coupling parameters β_K (for K=1,N), and the metal band width γ fix the self-energies and therefore the modifications to the molecular Green's function arising from the adsorption interaction. The imaginary part of the self-energy, Δ, also enter into the formula for the conductance (Eqs. 2,3).

The computation is then complete - the conductances are found from eqs. (2) and (3), with Hamiltonians and self-energies determined by Eqs. 5-8. The simple extended Huckel model is just

$$H^{(H)} = \sum h_{ij} a_{i\mu}^{+} a_{j\mu} \tag{9}$$

Here h_{ij} is the matrix element of the one electron Hamiltonian between basis orbitals i and j; i and j run over all the basis orbitals, while μ labels the spin. We employ the extended Huckel model, with basis orbitals non-orthogonal, and the matrix elements chosen according to a particular semiempirical recipe. We use standard extended Huckel parametrization in all calculations.[35b]

3. Molecular Wire Circuits: Chemical Effects

At the simplest level, we utilize the extended Huckel model[35] to characterize the electronic structure of the molecule, the Newns Anderson interaction[31] between molecule and electrode, and the simple linear molecular structure. This gives a Green's function as in that of Eq. 4, with self-energy terms as in Eq. 6.

Results for S-phenyl-S wire are shown in Fig. 2. It shows the square of the Green's function between site 1 and site N as a function of the injection energy at the electrodes (which becomes the effective Fermi energy - see comments in Section 7).

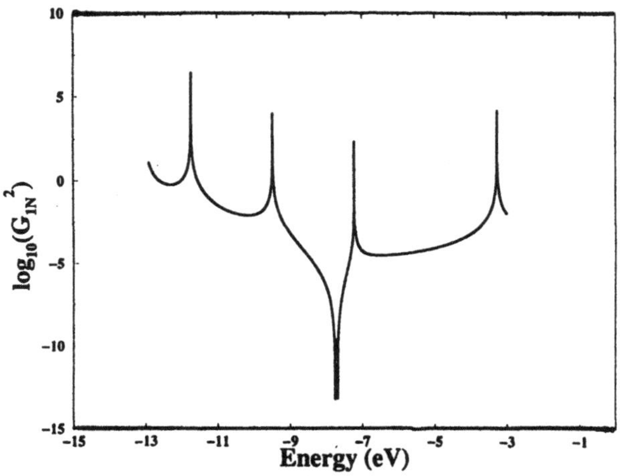

Fig. 2. Green's function as a function of the injection energy, showing resonances at the eigenvalues and cancellations at energies around -8.0 eV for S-C_6H_4-S.

Light vertical lines show the molecular orbital energy levels, while the heavy line is the Green's function squared. There are three relationships between features in the Green's function and features in the eigenvalue spectrum. In some cases, the two are entirely unrelated: that is, the Green's function passes smoothly through the position where molecular eigenvalues are. This essentially arises from overlap: in the lowest order, we can represent the Green's function as (with s labeling the molecular orbitals and Σ a self-energy)

$$G_{1N} = \sum_s <1s><sN>/(E - E_s + \Sigma) \tag{10}$$

The resonances that one would expect at positions where the injection energy is equal to the molecular eigenvalue, or (in the Huckel model) an orbital energy, may not occur if the overlap matrix element vanishes - that is, if the atomic orbital into which charge is injected is of different symmetry from the molecular orbital in question. For example, if injection occurred into a σ orbital, one would not expect resonant assisted tunneling when π orbitals cross the injection energy surface. This is exactly what occurs in Fig. 2: the orbitals for injection (at the terminal S atoms) are taken to be π structures, and all of the orbitals shown are of sigma type (so that the matrix elements of (4) vanish), except for those at -11.8, -9.4, -7.1 and -3.2eV.

There are features that appear to be minima in the Green's function. In fact, these are not minima but rather cases in which a near singularity, or resonance, has a negative residue - that is, the relative sign of the molecular orbital at the first and last sites of the bridge is negative. Then this term will correspond to the negative contribution to the conductance, which will dominate in the region of that resonance.

The richly structured conductance of Fig. 2 has not yet been directly observed, although room temperature coulomb blockade and molecular currents have been demonstrated for this molecule.[1] In the regions of resonant tunneling, the electron current moves essentially without scattering, since no inelastic scattering mechanisms have been included in our formalism. One would expect that under these conditions disorder scattering (see the next section) or vibrational scattering will reduce the rate.

Finally, it is useful to relate our behavior here to so-called electron type and hole-type superexchange reactions, often discussed[11,37,38,39] for long range electron transfer. When the injection energy is close to the lowest unoccupied molecular orbital (LUMO), that resonance will dominate the Green's function; in an electron-type superexchange picture, transport is mediated by the electrons. Similarly, when the injection energy lies closer to the highest occupied molecular orbital (HOMO) the analogue of hole type superexchange provides the current, with the resonance from the HOMO dominating the Green's function.

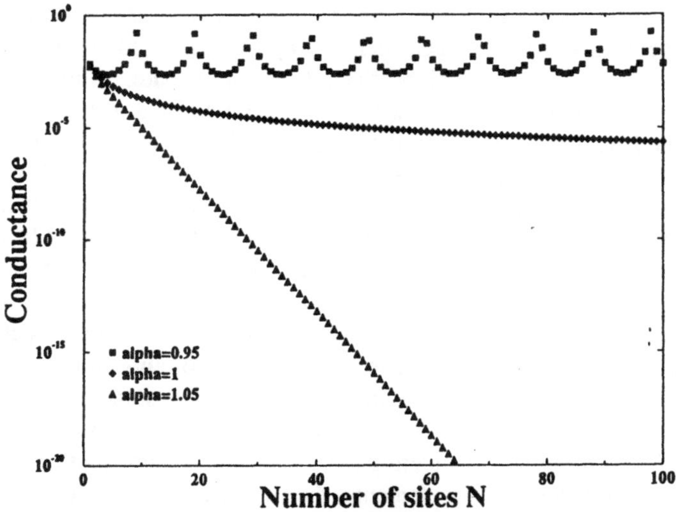

Fig. 3. Distance dependence of the conductance. Three kinds of dependences arise depending upon α. When $\alpha >> 1$, the conductance decreases exponentially with the number of bridge sites *N*. When $\alpha < 1$, the conductance exhibits smooth, bounded oscillations with *N*. When $\alpha \sim 1$, the conductance decreases as $1/N^2$. These curves were computed with $\beta_X/2\gamma = 0.1$. The conductance is measured in units of $e^2/\hbar$. From Ref. 21, *J.Chem.Phys.* 101, 6856 (1994). $\alpha = \varepsilon/2V$, ε being the site energy and V the site-site interaction in the wire.

In long range electron transfer, it is nearly always observed that rate constants decay exponentially with the distance between donor and acceptor.[7,9,11,37,38,40] This can be understood on several theoretical bases, including simple iteration of the equation for the effective matrix elements (Lippmann Schwinger equation).[38] When the orbital energies of the bridge between donor and acceptor become degenerate with donor or acceptor levels, however, neither donor nor acceptor is any longer well defined electronically, and a rate constant per se may not exist. This is not true of molecular wires, where the initial states and final states are fixed by the continuum in the electrodes, and the current can always be measured. One then might expect exponential decay of the conductance in situations where there is an energy gap between the injection level and molecular orbitals, but differing decay when injection occurs at an energy corresponding to an orbital eigenstate of the molecule. This is precisely what is computed:[21,41] Figure 3 shows the decay with distance for a chain of degenerate sites. When the injection energy is degenerate with a band of molecular orbitals, the conductance shows oscillations due to phase, but no overall decay with distance: this is true wire behavior, in which the current is carried very long distances without substantial loss. On the other hand, when the gap between injection energy and eigenvalues is nonvanishing, exponential decay is indeed observed, as in long range electron transfer. Finally, for the case where the injection is just at the band edge of occupied states, power law behavior is seen.

4. Effects of Electron Site Disorder

Considerations given thus far are for a molecular chain, and we assumed that all the atomic site energies are fixed. In general, one would not expect this to be true. Even in a formally homogeneous molecular chain such as polyethylene or polyacetylene, local site energetics and local bond strength might differ from site to site. This disorder could come from chemical functionalization such as the presence of a methyl group, from a local environmental perturbation due to the presence of other molecules or sections of molecules, from the distorted bonding geometry, or from strain effects. This suggests that both the off-diagonal and diagonal matrix elements of the Huckel Hamiltonian - that is, both the tunneling integrals between sites and the local site energies - could be disordered. This is highly reminiscent of the diagonal and off-diagonal disorder features that characterize[34] the Anderson localization in solid state studies. It is, therefore, of interest to see what effects this disorder might have on the conductance of a molecular wire.

Only a few studies of this kind have appeared.[22,24] An outstanding feature is that the statistical behavior of the conductance must now be studied, since the disorder effects are random and can only make sense in an ensemble of observations. The appropriate ensemble can be found simply by calculating the conductance at one particular realization of site energies and tunneling matrix elements, then repeating with further realizations. We confine our attention here to diagonal disorder in a simple linear chain molecule.

The most striking observation[24] is that the conductance distribution is of log normal type: that is, the probability of observing the logarithm of the conductance is distributed in a Gaussian fashion, but the conductance itself is not. This is important in characterizing the statistics and explaining transport in disordered systems.

Disorder substantially modifies the behavior of the conductance. One striking result[22] is shown in Fig. 4: here we plot the conductance as a function of disorder, for injection energies into a band of states and in the gap between molecular eigenvalues. Notice that for injection at an eigenvalue, where according to the results of the last section we would expect molecular wire behavior without exponential decay, the disorder causes localization of the states, and substantially decreases the conductance. On the other hand, when injection occurs into the gap, disorder actually <u>increases</u> the conductance. This can be understood in terms of stochastic modulation of the eigenvalues: occasionally, due to the presence of disorder, one of the molecular eigenvalues will become degenerate with the injection level, and a resonance assisted transfer will occur. This resonance assisted transfer permits increased current due to disorder (in fact, such stochastically modulated detection appears to be of use in biology, including signaling among crayfish!).

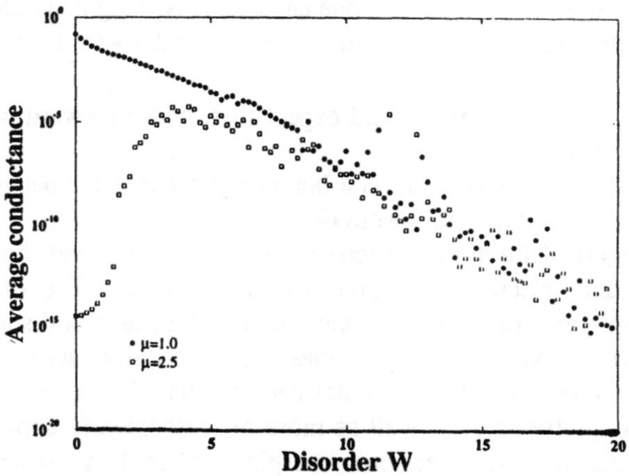

Fig. 4. Effect of disorder on the average conductance, for two values of the Fermi level energy. When the Fermi energy lies within the undisordered wire bandwidth (μ=1), the conductance decreases monotonically with disorder. In the opposite situation (μ=2.5), the conductance first increases until W=5, then decreases. The conductance is measured in units of $e^2/\hbar$. The parameters of the wire are N=20 and Δ=0.1. From Ref. 22, *J.Chem.Phys.* **101**, 5172 (1994).

5. Electron Repulsion Effects: Coulomb and Eigenvalue Ladders

In quantum dot structures, single electron tunneling phenomena have been under intense investigation both theoretically and experimentally.[17,33,42,43] Both the coulomb blockade (prevention of current through a quantum dot until an injection barrier has been overcome), and coulomb staircase (regular jumps in the current as integral numbers of electron are deposited on the dot) have been well studied in quantum dot structures, although because of their characteristic sizes, these are seen only at very low temperature.

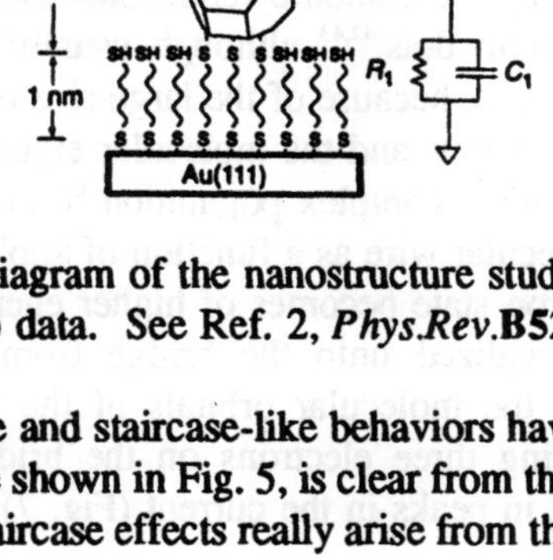

Fig. 5. (a) Schematic diagram of the nanostructure studied. (b) Double junction model used to explain the I(V) data. See Ref. 2, *Phys.Rev.***B52**, 9071 (1995).

While blockade and staircase-like behaviors have been observed[2] for the sort of molecular wire structure shown in Fig. 5, is clear from the equivalent circuit written there that the blockade and staircase effects really arise from the gold nanodot, rather than from the molecular wire itself. Nevertheless, one can certainly imagine coulomb blockade and molecular eigenvalue staircase phenomena on the molecular wire itself, in a situation like that in Fig. 1. If $\frac{\beta_1}{\beta_N} >> 1$, one would expect that, under forward bias, charge would build up on the molecule, since there is a bottleneck for transfer between the downfield end of the molecule and the metal electrode.

Investigation of these phenomena requires going beyond the extended Huckel model, and including electron repulsion within the wire. The explanation usually given[33,42,43] for single electron tunneling and coulomb blockade and staircase behavior in mesoscopic systems is based on a quasi-classical description involving capacitance at interfaces. While molecular capacitance has been discussed in several situations,[44,45] it is probably more straightforward, as well as more microscopically approachable, to use a hamiltonian formula in which electronic energies, rather than capacitances, enter; the discussions are in fact probably equivalent.

As a first attempt to understand molecular coulomb blockades, we have studied[25] a four site wire characterized by a Hubbard Hamiltonian - that is, by an extended Huckel Hamiltonian in which one also includes repulsive energies of interaction between the up and down spin electrons at each site. That is, one writes the Hubbard Hamiltonian as

$$H_{Hub} = H_{Huc} + \sum_i n_{i\uparrow} n_{i\downarrow} U \tag{11}$$

Here U is the Hubbard repulsion parameter, while $n_{i\uparrow}$ is the population of spin up electrons in the ith atomic basis orbital.

Once the Hubbard repulsions are included, one can no longer simply solve the Green's function using the eigenvalues of the Huckel Hamiltonian, since an exact solution does not really exist. We have thus used the Hartree-Fock approximation, to get a qualitative understanding of molecular repulsion effects. In the Hartree-Fock approximation, one replaces, in the Hamiltonian for the up spin electrons, the population of down spin electrons by its average.

The coulomb repulsion produces a blockade-like feature at low bias - that is, no electrons can be forced onto the molecule until a molecular level becomes available to them; the onset of this molecular level depends both on the molecular eigenvalue staircase (that is, on the orbital energies in the absence of coulomb repulsion) and on the changes in that staircase brought about by the coulomb repulsion. Similar situations have been studied in the context of quantum dots,[42,43] although usually[46] the discreteness of the eigenvalue spectrum is unimportant, because of the large size of the quantum dot; this is definitely not true of molecular wires, and the molecular eigenvalue staircase combines with the coulomb repulsion to yield complex population behaviors.[25] Fig. 6 shows the population of the four-site molecular wire as a function of applied field: notice that at a particular onset, the four electron state becomes of higher energy than the five electron state, and an electron is delocalized onto the bridge from the injecting electrode. Similarly, electrons can leave the molecular orbitals of the bridge and go on to the accepting electrode, thus leaving three electrons on the bridge at one point. These injection and loss events result in peaks in the current (Fig. 7)

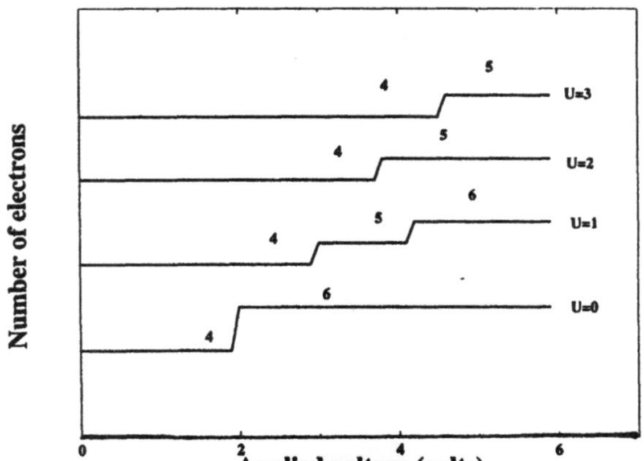

Fig. 6. Dependence of charging effects on Hubbard U. Number of electrons of the most stable configuration versus applied voltage. In the absence of repulsion (U=O), the number of electrons increases by 2 when the LUMO crosses the donor Fermi level (V=2V). For U>O, the number of electronics increases by 1 only because of repulsion. From Ref. 25, *J.Chem.Phys.* **104**, 7296 (1996).

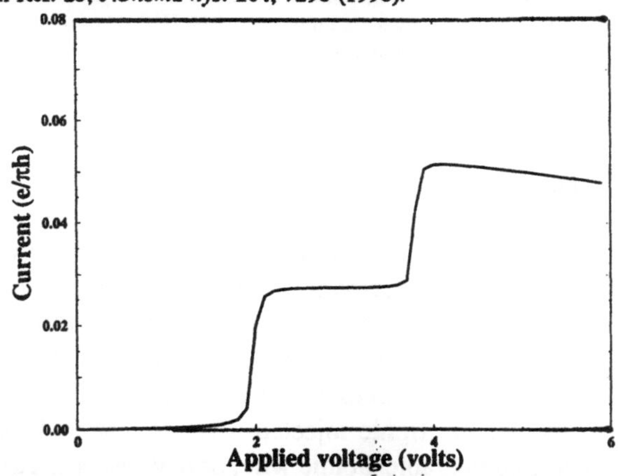

Fig. 7. Current-voltage characteristics of a molecular wire. The characteristic is very stairway shaped: current jumps occur at V~2 and V~4 volts, and current plateaus in-between. Parameters: N=4, β=-2.4 eV, U=2 eV, Fermi level-LUMO gap=1 eV, $\beta_1=\beta_N$=-0.5 eV, γ=10 eV. From Ref. 25, *J.Chem.Phys.* **104**, 7296 (1996).

There are several things to note about these currents: first, the jumps are no longer evenly spaced as they are in the standard mesoscopic coulomb staircase; this is the effect of the molecular eigenvalue. Second, notice that the potentials at which the steps occur are very broadly spaced - with these potential spacings, one would expect these staircase effects to occur at room temperature, and indeed precisely such effects have been seen at room temperature in molecular wire systems (although, again, there the blockade is probably[2] on the nanodot rather than on the wire itself).

Finally, rectification phenomena can be clearly observed in molecular wire structures.[3,47] If the wire is inhomogeneous, it will almost certainly be easier to inject electrons from one electrode than the other, and therefore one expects asymmetry in the current. This is shown in Fig. 8, and similar rectifications have been observed several times in molecular wire structures. Prediction of such rectification behaviors occurred long before experimental observation[48], and this observation marks a significant milestone in the understanding and characterization of molecular electronics.

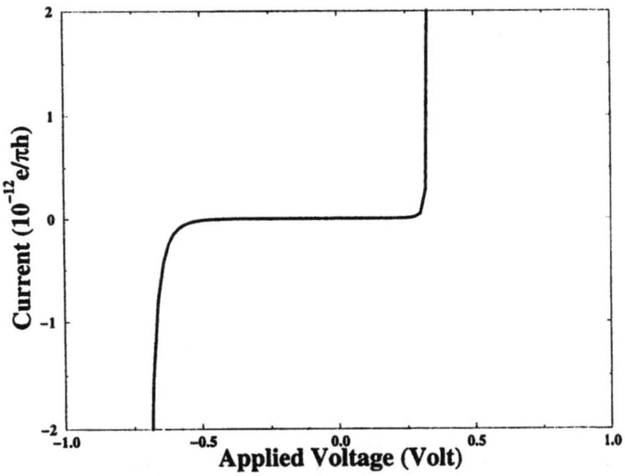

Fig. 8. Current-voltage characteristics of a heterogenous wire. The I-V is asymmetric and shows rectifying behavior. The wire consists of two segments, each six units long. The HOMO of the left segment lies 0.2 eV below the Fermi level, and the HOMO of the right segment 1 eV below. Parameters: β=-0.25 eV, U=0 eV, $\beta_1=\beta_N$=-0.5 eV, γ=1.0 eV. From Ref. 25, *J.Chem.Phys.* **104**, 7296 (1996).

6. Experimental Comparisons

Use of both break junction circuits and STM spectroscopy have resulted in reported measurements of molecular wire conductances.[1-4] Table I shows a few of the calculated conductances, and, in some cases, computational results. There are three striking features: first, the conductances are substantially less than the quantum of conductance, $[12.9\ k\Omega]^{-1}$. This can be understood from the Landauer formula[5] or from the simple physics: resistance is scattering, and the electronic injected wave is scattered more successfully with a molecular wire than a comparable atomic wire, or macroscopic circuit. Indeed, the transmission coefficients in the Landauer formula are generally less than 10^{-3} for molecular wires. Second, the observed conductances of molecular wires result in substantial currents: for an applied field of one volt, a characteristic molecular wire conductance of $(10M\Omega)^{-1}$ will result in a current of 100nA, easily measured and electronically useful. Third, the agreement between the very simple theoretical model of Eq. (2) and the reported experimental data is surprisingly close.

Because measurement of conductance in molecular circuits is really still so new, a large number of the predictions of theory have not yet been tested: these include length and temperature dependence of the conductance, measurements with electrodes of different metals, observations of vibrational structure, bonding dependencies and statistical aspects of the conduction.

TABLE 1. The Comparison of Calculated and Measured Resistances[a)]

System	Calculated Resistance (MΩ)	Experimental Results (MΩ)
gold-sulfur-aryl-sulfur-gold	.03[a)]	13.3~22[b)]
gold-sulfur-xylyl-sulfur-gold	9.2	6~9[c)]
gold-C_{60}-tip	116.6[d)]	54.8±13[e)]

a) Using a value of $\beta_1 = \beta_n = -3.4$ eV
b) ref. 1, 6
c) ref. 2
d) Using a value of $\beta_1 = \beta_n = -0.47$ eV
e) ref. 4

7. Speculations

Actual circuit control elements in contemporary solid-state work usually involve a 3-terminal device. Figure 9 shows a three terminal FET-type structure in which the quinone species (the ring with the two oxygens) acts as an external capacitive gate on the molecular wire: that is, the electrons will be concentrated on the oxygens, and the coulomb repulsion arising from the presence of the electrons will modify the molecular orbital energy structures in the molecular wire, and therefore modify the conductance of the wire itself. FET electrode circuits of this kind, based on molecular structures, can be prepared, and should be studied as part of the next chapter in molecular electronics.

Use of the molecular eigenvalue staircase as a switch or transistor structure is attractive, because the large spacings (as opposed to much smaller, meV scale steps available in the coulomb blockade in typical quantum dots) should permit room temperature function. Some speculations, calculations, and early measurements of analogous phenomena in small scale quantum dots are available,[33,42,43] and some measurements are beginning to appear on molecular scale structures.[1-4]

Additional topics whose theoretical analysis remains incomplete, but whose experimental manifestations are important, include vibrational effects (Franck-Condon type sidebands on the conductance spectrum, or (more interestingly), possible quasi-particle transport due to dressed particles of polaron or soliton type, control of the injection energy by equalization of Fermi potentials in the wire structure and the electrodes themselves, surface bonding effects and symmetry breaking due to differing surface interactions, and effects of fluctuations and statistics at the single molecule level.

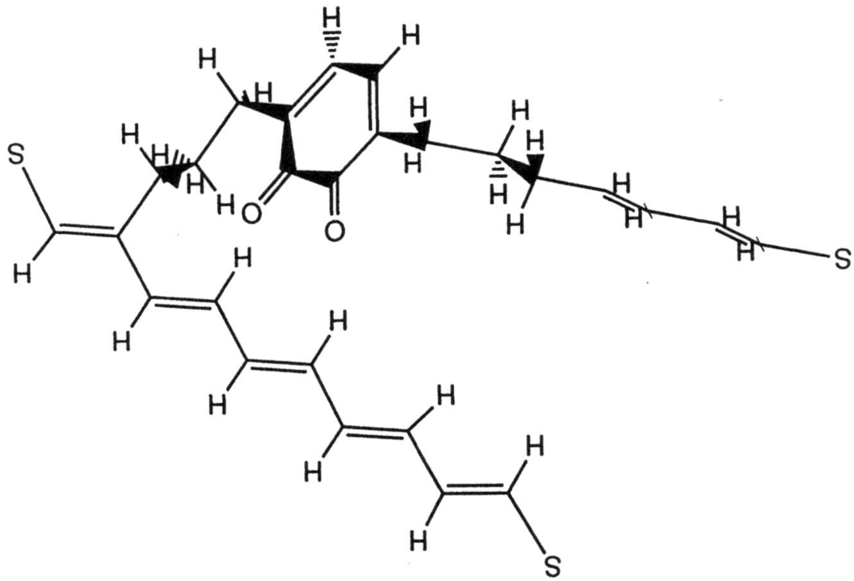

Fig. 9. A molecular FET-type structure. Dark circles correspond to carbon atoms.

Acknowledgements:

We are grateful to the Chemistry Division of the NST and ONR for ongoing support of this work; we thank J. Jortner, R. VanDuyne, M. Wasielewski and M. Reed for very instructive discussions. VM and MR are grateful to NSF and CONICIT for support through the collaborative Venezuela/U.S. research program.

References:

1. M.A. Reed, C. Zhou, C.J. Muller, T.P. Burgin and J.M. Tour, submitted to *Science*.
2. R.P. Andres, T. Bein, M. Dorogi, S. Feng, J.I. Henderson, C.P. Kubiak, W. Mahoney, R.G. Osifchin and R. Reifenberger, *Science* **272**, 1323 (1996); M. Dorogi, J. Gomez, R. Osifchin, R.P. Andres and R. Reifenberger, *Phys. Rev.* **B 52**, 9071 (1995).
3. C.M. Fisher, M. Burghard, S. Roth and K.v. Klitzing, *Appl.Phys.Lett.* **66**, 3331 (1995); also N.J. Tao, *Phys.Rev.Lett.* **76**, 4066 (1996) and S. Roth, M. Burghardt and C.M. Fischer, in Ref. 7.
4. C. Joachim, J.K. Gimzewski, R.R. Schlittler and C. Chavy, *Phys.Rev.Lett.* **74**, 2102 (1995).
5. R. Landauer, *Philos.Mag.* **21**, 863 (1970); R. Landauer, *IBM J. Res. Dev.* **1**, 223 (1957); R. Landauer, in *Localization, Interaction and Transport Phenomena*, edited by B. Kramer, G. Bergmann, and Y. Bruynseraede (Springer, New York, 1985), pp. 38-50.
6. C. Zhou, C.J. Muller, M.A. Reed, T.P. Burgin and J.M. Tour, in Ref. 7.
7. M.A. Ratner and J. Jortner, eds., *Molecular Electronics*, (Butterworth, 1997).
8. (a) D. de Vault, *Quantum-Mechanical Tunnelling in Biological Systems*, 2nd edn., (Cambridge University Press, Cambridge, 1984); (b) R.A. Marcus and N. Sutin, *Biochem. Biophys. Acta*, **811**, 265 (1985); (c) M.D. Newton and N. Sutin, *Annu.Rev.Phys.Chem.*, **35**, 437 (1985); (d) K.V. Mikkelsen and M.A. Ratner, *Chem. Rev.*, **87**, 113 (1987); (e) J. Jortner and B. Pullman (eds.), *Proc. 22nd Jerusalem Symposium*, (Reidel, Dordrecht, 1989).

9. *J. Photochem.Photobiol.* **82**, #1-3, ed. L.G. Arnaut (1994).
10. *Chemical Physics* **176**, #2,3 (1993).
11. R.A. Marcus, *Revs. Mod. Phys.* **65**, 599 (1993).
12. G.L. Closs and J.R. Miller, *Science* **240**, 440 (1988).
13. Ch. Joachim and P. Sautet, in *Scanning Tunneling Microscopy and Related Methods*, edited by R.J. Behm, N. Garcia, and H. Rohrer (Kluwer Academic, Dordrecht, 1990); P. Sautet and Ch. Joachim, *Chem. Phys. Lett.* **185**, 23 (1991).
14. C. Joachim and J.F. Vinussa, *Europhys. Lett.* **33**, 635 (1996); C. Joachim and J.K. Gimzewski, *Europhys. Lett.* **30**, 409 (1995).
15. M. Sumetskii, *Phys.Rev.* **B48**, 4586 (1993); M. Sumetskii, *J. Phys. Cond. Matt.* **3**, 2651 (1991).
16. M. Sumetskii, *Sov. Phys. JETP* **62**, 355 (1985).
17. S. Datta, *Electronic Transport in Mesoscopic Systems* (Cambridge, 1995).
18. M.P. Samanta, W. Tian, S. Datta, J.I. Henderson and C.P. Kubiak, *Phys. Rev.* **B53**, 7676 (1996).
19. N.D. Lang, *Phys. Rev.* **B52**, 5335 (1995); A. Yazdani, D.M. Eigler and N.D. Lang, *Science* **272**, 1921 (1996).
20. V. Mujica, M. Kemp and M.A. Ratner, *J. Chem. Phys.* **101**, 6849 (1994).
21. V. Mujica, M. Kemp and M.A. Ratner, *J. Chem. Phys.* **101**, 6856 (1994).
22. M. Kemp, V. Mujica and M.A. Ratner, *J. Chem. Phys.* **101**, 5172 (1994).
23. R.P. Van Duyne and M. Smith (unpublished).
24. M. Kemp, A. Roitberg, V. Mujica, T. Wanta and M.A. Ratner, *J. Phys. Chem.* **100**, 8349 (1996).
25. V. Mujica, M. Kemp, A. Roitberg and M.A. Ratner, in *Condensed Matter Theories 11*, edited by E.V. Ludena, P. Vashishta and R.F. Bishop (Nova Science, Commack, 1996) (in press); V. Mujica, M. Kemp, A. Roitberg and M. Ratner, *J.Chem.Phys.* **104**, 7296 (1996).
26. M. Kemp, Y. Mao, V. Mujica, A. Roitberg and M.A. Ratner, submitted to *Science*.
27. V. Mujica and G. Doyen, *Int. J. Quantum Chem. S* **27**, 687 (1993).
28. H. Ou-Yang, R.A. Marcus and B. Källebring, *J. Chem. Phys.* **100**, 7814 (1994).
29. J.P. Reimers and N.S. Hush, *J. Photochem. Photobiol.* **82**, 31 (1994).
30. J. Bardeen, *Phys. Rev. Lett.* **6**, 57 (1961).
31. D.M. Newns, *Phys. Rev.* **178**, 1123 (1969); P.W. Anderson, *Phys. Rev.* **124**, 41 (1961).
32. J. Tersoff and D.R. Hamann, *Phys. Rev.* **B31**, 805 (1985).
33. A.N. Korotkov, in ref. 7.
34. P.W. Anderson, *Phys. Rev.* **109**, 1492 (1958).
35. R. Hoffmann in *J. Chem. Phys.* **39**, 1397 (1963).
36. J.M. Krans, C.J. Muller, N. van der Post, F.R. Postma, A.P. Sutton, T.N. Todorov and J.M. Ruitenbeek, *Phys. Rev. Lett.* **74**, 2146 (1995).
37. H.M. McConnell, *J. Chem. Phys.* **35**, 508 (1961).
38. M.A. Ratner, *J. Phys. Chem.* **94**, 4877 (1990).
39. B.P. Paulson, L.A. Curtiss, B. Bal, G.L. Closs and J.R. Miller, *J. Am. Chem. Soc.* **118**, 378 (1996).
40. P. Barbara, T.J. Meyer and M.A. Ratner, *J. Phys. Chem.*, **100**, 13148 (1996).
41. J. Evenson and M. Karplus, *J. Chem. Phys.* **96**, 5272 (1992).
42. *Single Charge Tunneling. Coulomb Blockade Phenomena in Nanostructures*, edited by M. Grabert and M.H. Devoret (Plenum, New York, 1992).
43. *Mesoscopic Phenomena in Solids*, edited by B.L. Altschuler, P.A. Lee, and R.A. Webb (Elsevier, Amsterdam, 1991); *Nanostructures and Mesoscopic Systems*, edited by W.P. Kirk and M.A. Reed (Academic, New York, 1991); A.N. Korotkov, D.V. Averin and K.K. Likharev, *Physica B* **165**, 927 (1990).
44. F.T. Smith, *J. Chem. Phys.* **34**, 793 (1961).
45. L. Belkhir, *Phys. Rev.* **B50**, 8885 (1994).
46. But not always. Compare ref. 37, also B.Su, V.J. Goldman and J.E. Cunningham, *Phys. Rev.* **B46**, 7644 (1990).
47. A.S. Martin, J.R. Sambles and G. Ashwell. *Phys. Rev. Lett.* **70**, 218 (1994).
48. A. Aviram and M.A. Ratner, *Chem. Phys. Lett.* **29**, 277 (1974).
49. D.L. Klein, P.L. McEuen, J.E.F. Katari, R. Roth and A.P. Alivisatos, *Appl. Phys. Lett.* **68**, 2574 (1996).

TUNNEL TRANSPORT THROUGH A MOLECULAR WIRE

M. MAGOGA AND C. JOACHIM
CEMES-LOE
29, rue Jeanne Marvig, 31000 Toulouse, France

1. Introduction.

Let us consider an electric circuit made of two electrodes closed by a single molecule. Adsorbed at both ends to the electrodes surface, this molecule creates an electronic coupling between them [1, 2] and becomes a component of the circuit whose properties are usually followed by measuring current-voltage $I(V)$ characteristics [3, 4]. Such measurements were for example performed on a single C_{60} molecule [5]. It results a linear $I(V)$ characteristic for bias voltage lower than the C_{60} homo-lumo gap (Figure 1).

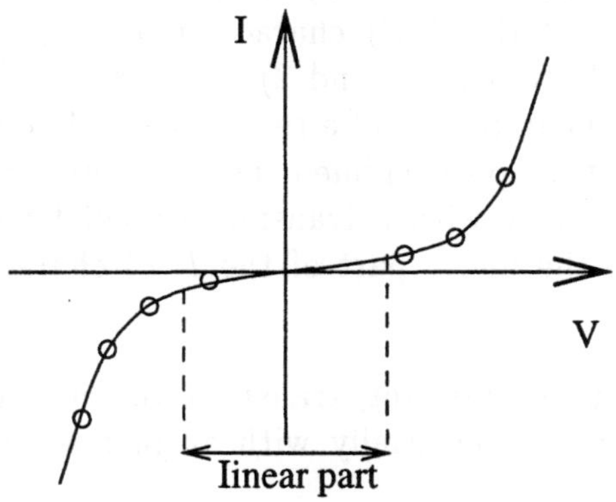

Figure 1. Typical $I(V)$ characteristic through a molecule.

Considering the Landauer formula relating the current intensity in the circuit to the electronic transparency of the molecule, two regimes of transport can be distinguished by looking at the electronic transmission spectrum[6, 7, 8] i.e. the mono-electronic transmission coefficient through the molecule plotted as a function of the electron incident energy (Figure 2).

C. Joachim and S. Roth (eds.), Atomic and Molecular Wires, 219-227.

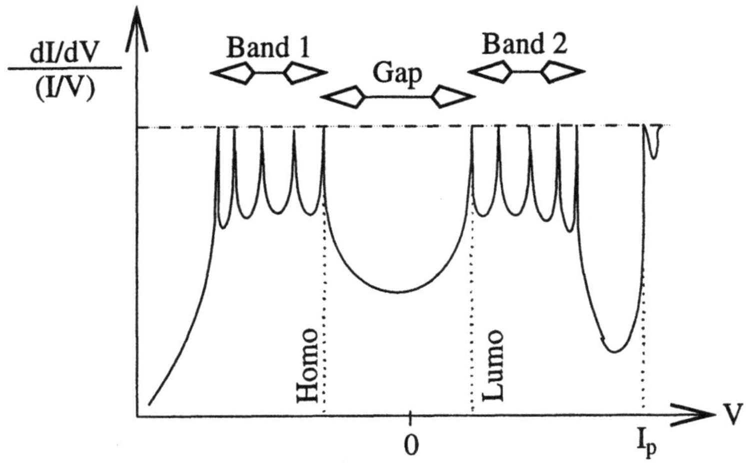

Figure 2. An example of an electronic transmission spectrum through a molecule (I_p is the ionization potential).

1. In the energy ranges containing the peaks, which corresponds to the non-linear part of the $I(V)$ characteristic (Fig. 1), the conduction (band 1) and the valence (band 2) bands are in formation separated by a gap. Each peak results of a resonance with a wire molecular level.
2. In the gap, the transport regime is non-resonant. Electrons are tunneling through the wire with a transmission coefficient $0 < t < 1$. This corresponds to the linear part of the $I(V)$ characteristic (Fig. 1), at small voltage.

In case 2, since it is a tunneling transport regime, the transmission coefficient will decrease exponentially with an increase of the length of the molecule L [9] :

$$t(L) = t_0 e^{-\gamma L} \tag{1}$$

Clearly a wire is more useful if its electronic properties are length independent but this implies a non linear I-V characteristic which is useless too. Therefore, for a gaped molecule a choice must be made between linearity and conduction. Furthermore, electrons are tunneling through the molecule which may lead to a better control of t_0 and γ than a non-structured tunnel barrier. Thus, we have chosen to keep the linearity and to explore how γ in (1) can be made small, by optimizing the molecular wire chemical structure.

In section 2, a tight-binding model is presented to show how γ is controlled by the wire internal structure. In section 3 electronic transmission spectra calculated by the ESQC technique are discussed for a typical oligomer, the (oligo(thiophene ethynylene), presented as a good molecular wire in the literature [10].

2. Tight-binding theory of the tunnel slope.

To study the general trends of a electrode-wire-electrode system, a one-dimensional tight-binding model can be chosen. Electrodes are represented by two infinite chains and the molecule is a regular chain of identical units, each one containing two effective levels (Figure 3).

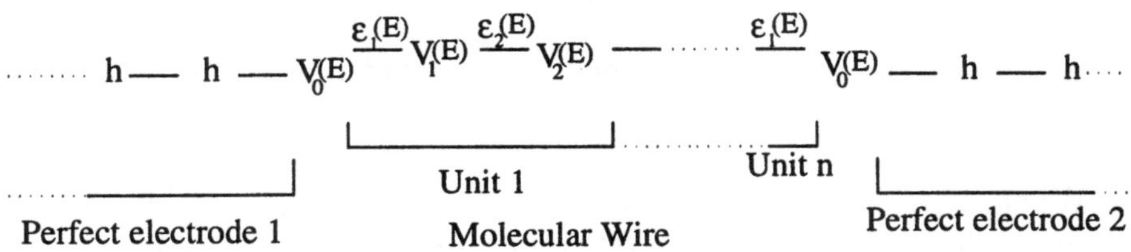

Figure 3. The tight-binding model of the electrode-wire-electrode system .

In a periodic unit, $\epsilon_1(E)$ and $\epsilon_2(E)$ are the energy levels, $V_1(E)$ and $V_2(E)$ the interaction coefficients calculated from a Löwdin effective Hamiltonian [1]. In the electrodes h is the hopping integral and e, the energy level of each site, is taken equal to zero, representing a semi-filled band system with a Fermi energy $E_f = 0.0$. Finally, $V_0(E)$ is the interaction between the electrodes and the wire.

The Bloch wave propagation in the chain through the molecule is studied using the transfer matrix technique [11], $T(E)$ is a T-matrix written (assuming n is the number of unit)

$$T(E) = A(E)(M_1(E)M_2(E))^n B(E) \tag{2}$$

with $A(E)$ the electrode-wire T-matrix, $B(E)$ the wire-electrode T-matrix and $M_1(E)$ and $M_2(E)$ the T-Matrix for the periodic unit of the wire.

Let λ_+ and λ_- be the eigenvalues of the matrix $M(E) = M_1(E)M_2(E)$. As $\det(M(E)) = 1$, it comes $\lambda_+\lambda_- = 1$. The transmission coefficient $t(E)$ can be written in term of these eigenvalues, with a_{ij} (resp. b_{ij}) the matrix elements of A (resp. B):

$$t(E) = \frac{1}{|T_{11}|^2} = \frac{1}{|a_{11}b_{11}\lambda_+{}^n + a_{12}b_{21}\lambda_-{}^n|^2} \tag{3}$$

Assuming that λ_+ is the highest, $t(E)$ can be rewritten for large n :

$$t(E) \simeq t_0(a_{11}, b_{11})e^{-2n\log(|\lambda_+|)} \tag{4}$$

Therefore the slope $\gamma = 2\log(|\lambda_+|)$ of $t(E)$ is controlled by the highest $M(E)$'s eigenvalue. But the characteristic equation of $M(E)$ is :

$$\lambda^2 - Tr(M(E))\lambda + 1 = 0 \tag{5}$$

where discriminant is

$$\Delta(E) = Tr(M(E))^2 - 4 \tag{6}$$

Then, the wire band structure can be located depending on the $\Delta(E)$ sign. When $\Delta(E) < 0$, the eigenvalues are complex. Since $\lambda_+\lambda_- = 1$, it comes $|\lambda_+| = 1$, the corresponding energy value belong to a band of the molecular wire. When $\Delta(E) > 0$ the eigenvalues are real and $|\lambda_+|$ can be different from unity, the corresponding energy values are located in a gap.

In the general case, $\deg(\Delta(E)) = 2p > 2$ and $\Delta(E)$ can be written as a product of its roots (see figure 4):

$$\Delta(E) = \prod_{i=1}^{2p}(E - E_i)(E + E_i) \tag{7}$$

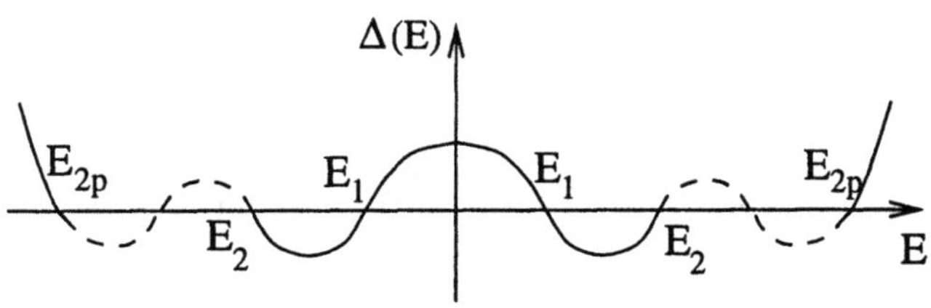

Figure 4. Representation of the variation of $\Delta(E)$ in the general case where $\deg(\Delta(E)) = 2p$.

Let us write the gap $\chi = 2E_1$ and $\Delta(0) = (E_1)^2\prod_{i=2}^{2p} E_i^2 > 0$. Then, the eigenvalues $\lambda_\pm$ can be written as:

$$\lambda_\pm(0) = \frac{\chi\prod_{i=2}^{2p} E_i \pm \sqrt{\chi\prod_{i=2}^{2p} E_i^2 + 16}}{4} \tag{8}$$

As expected, the slope varies with the square root of the molecular wire gap width [6]. Moreover, the analytic expression of $\lambda_\pm$ (i.e. of γ) shows also

a dependence on the product of all the roots. This proves that not only the gap of the wire controls the slope γ. Therefore, for a given gap, the optimization of the internal electronic structure of the wire may open the way to control also γ.
As an example, let us compare the case of a $H_2 - C(CH)_nC - H_2$ polyene chain with the case of an inorganic $H_2N - (SN)_n - NH_2$ oligomer. The simplest tight-binding representation of the polyene is :

$$\begin{cases} V_1(E) = \alpha \\ V_2(E) = \alpha' \\ \epsilon_1(E) = \epsilon_2(E) \end{cases} \tag{9}$$

In this case, it comes from (6) $\deg(\Delta(E)) = 4$, and from(8) with $y = \frac{\alpha}{\alpha'}$:

$$\lambda(0) = \frac{\chi(1+\frac{1}{y}) + \sqrt{\chi^2((1+\frac{1}{y})^2 + 16}}{4} \tag{10}$$

Therefore, assuming a gap $\chi = 1\,eV$ and $1 + \frac{1}{y} = 2.25$, the calculation of the slope from (10) gives $\gamma = 2\log(|\lambda_+|) = 0.574$.

For the $H_2N - (SN)_n - NH_2$ oligomer, the simplest tight-binding representation is

$$\begin{cases} V_1(E) = V_2(E) \\ \epsilon_1(E) = a \\ \epsilon_2(E) = a' \end{cases} \tag{11}$$

If now we put $\bar{a} = \frac{a}{a'}$, it comes from (8):

$$\lambda(0) = \frac{\chi\sqrt{\frac{\bar{a}^2+16}{4}} + \sqrt{\chi^2\sqrt{\frac{\bar{a}^2+16}{4}}^2 + 16}}{4} \tag{12}$$

Therefore, with again the same gap $\chi = 1eV$ and $\sqrt{\frac{\bar{a}^2+16}{4}} = 2.009$, the calculation gives $\gamma = 2\log(|\lambda_+|) = 0.498$.

This comparison confirms that not only the homo-lumo gap of a wire controls the tunneling slope γ. It's internal electronic structure, here represented by the repartition of its molecular levels in the energy spectra will also be important to reach a small γ. But notice that for two electronic structures as different as $H_2 - C(CH)_nC - H_2$ and $H_2N - (SN)_n - NH_2$, the two predicted γ are very close. This indicates either that the γ optimization will be hard or that the unit cell of these wires are much too simple.

3. Calculation.

Our tight-binding calculations have proved that the tunnel slope is homo-lumo dependent, but also a function of the complete electronic structure of the molecular wire. Therefore, to understand how to control γ, ESQC calculations were performed on oligomers proposed to be good molecular wires in the literature. The advantage of ESQC[12] is that it can be viewed as a direct extension of our previous tight-binding calculations. the rules explored may then be still valid for a calculation technique describing better the internal electronic structure of the molecule.

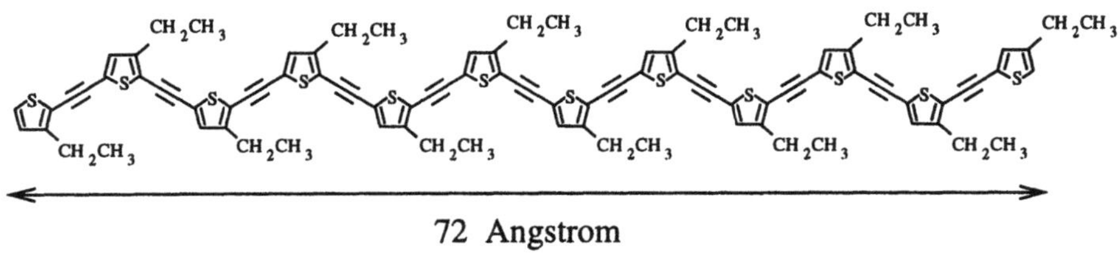

Figure 5. A 12 rings oligo(thiophene ethynylene).

All the molecular wires studied are adsorbed on gold pads. Each pad is described by its complete crystalline structure with a (100) surface for the molecule end adsorption. Each molecule is described by its full valence orbital structure. The extended Huckel approximation was used in ESQC due to the large electronic systems studied here (molecular wire of more than 200 atoms adsorbed on pads composed of 384 Au atoms were calculated).

As an example, let us show the behavior of the oligo(thiophene ethynylene)s [10] (Figure 5) which present the best slope γ for a gap at infinite length $\chi = 0.961\,eV$. The transmission spectra of this molecular wire are presented in Figure 5 for two different lengths (figure 6).

With these spectra, calculated for increasing length from 46 Å (nine thiophene units) to 90 Å (sixteen thiophene units), the slope γ can be deduced from the minimum values of $t(E)$ in the gap:

$$\gamma = 0.182\,\text{Å}^{-1}$$

Using the Landauer formula, the current intensity I in the circuit gap can be calculated for a bias voltage $V = 0.1\,V$, leading to I = 131.5 fA for the small one, I = 13.4 fA for the eleven units, and 0.309 fA for a usable length of 80 Å. Thus, for now, the best known wire provides a too low current to be measurable. Fortunately, another parameter is as important as the slope γ: the coefficient t_0 in (1). Indeed, if the value of this parameter

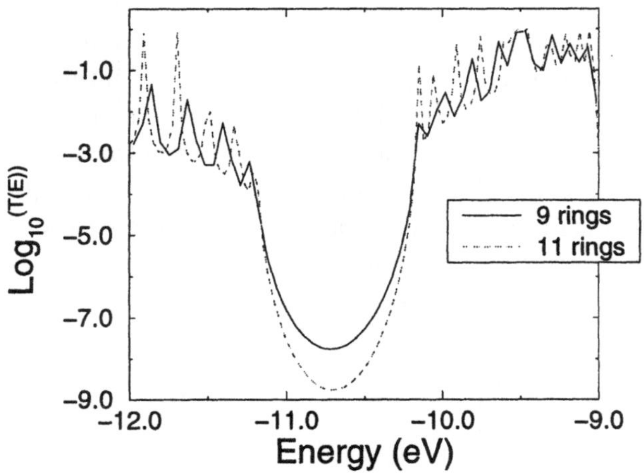

Figure 6. The oligo(thiophene ethynylene) spectra. Each resonance correspond to a molecular level.

is sufficiently high, then the current intensity will be measurable even for a 80 Å long oligomer. We now will prove the possibility to modify the value of t_0 without modifying γ.

For this purpose, another type of calculation was made with a different oligomer series [13], the $CH_2(CH)_nCH_2$ polyene, adsorbed on the gold pads either chemisorbed or physisorbed, leading to the transmission spectra presented in Figure 7.

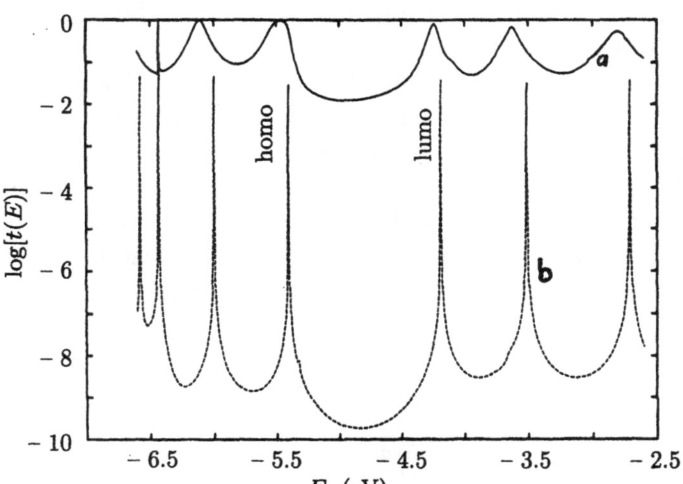

Figure 7. The transmission spectra of a chemisorbed and physisorbed polyene.

For the chemisorbed case (a) the resonances occurs with the molecular levels of symmetry π while in the second the ones of symmetry σ. The peaks are larger in the chemisorbed state showing that the interaction between molecular π levels and the electrodes continuum are stronger. This shows that by controlling the adsorption state, the prefactor t_0 in (1) can be well controlled. For the polyene case, going from the physisorbed to the chemisorbed state lead to a 5 order of magnitude increase of the average transmission coefficient without modifying the slope γ.

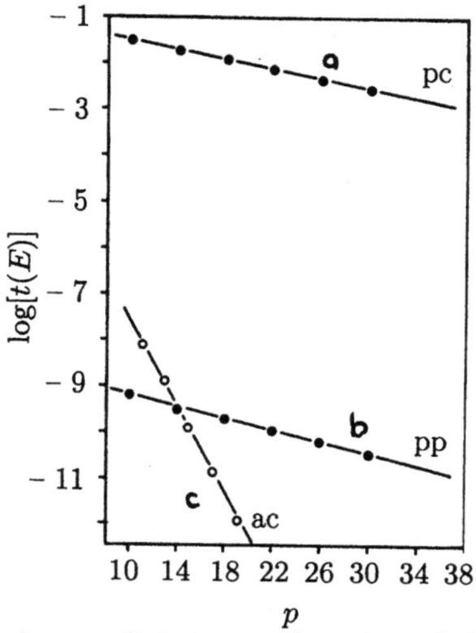

Figure 8. The transmission coefficient as a function of the unit number for polyenes and alkanes chains.

Finally, the figure (8) shows the length dependence of the lowest transmission coefficient taken in the gap for (a) chemisorbed (b) physisorbed polyene chains and (c) alkanes.

As expected, the electron delocalisation in a polyene leads to a better γ slope than for alkane. Thus, for a given internal molecular structure, it is possible to modify t_0 without any changes for γ. Modifications of the wire end, but not to the regular units along the oligomer, will increase (or decrease) the electrode-wire interactions and then t_0 but will not modify the characteristics of the wire. It's then possible to conceive independently the wire and the units that will connect it to the electrodes to optimize both t_0 and γ.

References

1. Joachim C. (1991) *New Journ. Chem.* **15**, 223.
2. Mujika V., Kemp M. and Ratner M. (1994) Electron conduction in molecular wires. 1. a scattering formalism., *J. Chem. Phys.* **101**, 6849–6855.
3. Joachim C. and Gimzewski J.K. (1995) Analysis of low-voltage i(v) characteristics of a single c_{60} molecule., *Europhysics Letters* **30**, 409–414.

4. Bumm L.A., Arnold J.J., Cygan M.T., Dunbar T.D., P. B.T., Jones II L., Allara D.L., Tour J.M. and Weiss P.S. (1996) Are single molecular wires conducting ?, *Science* **271**, 1705–1707.
5. Joachim C., Gimzewski J.K., Schlittler R.R. and Chavy C. (1995) Electronic transparence of a single c_{60} molecule., *Physical Review Letter* **74**, 2102–2105.
6. Samanta M.P., Tian W. and Datta S. (1996) Electronic conduction through organic molecules, *Physical Review B* **53**, R7626–R7629.
7. Mujica V., Kemp M. and Ratner M.A. (1994) Electron conduction in molecular wires. 2. application to scanning tunneling microscopy, *Journal of chemical physics* **101**, 6856–6864.
8. Joachim C. and Sautet P. (1990) *Scanning Tunneling Microscopy and related methods*, Kluwer Academic, Dordrecht.
9. Binnig G., Rohrer H. and Gerber C. (1982) *Helv. Phys. Acta* **55**, 726.
10. Pearson D., Schumm J. and Tour J. (1994) *Macromolecules* **27**, 2348.
11. Ziman J. (1979) *Models of disorder*, Cambridge University Press, London.
12. Sautet P. and Joachim C. (1988) Electronic interference produced by a benzene embedded in a polyacetylene chain, *Chem. Phys. Lett.* **153**, 511–516.
13. Joachim C. and Vinuesa J.F. (1996) Length dependence of the electronic transparence (conductance) of a molecular wire, *Europhysics Letters* **33**, 635.

SUBJECT INDEX

MIX
Papier aus verantwortungsvollen Quellen
Paper from responsible sources
FSC® C105338

If you have any concerns about our products,
you can contact us on
ProductSafety@springernature.com

In case Publisher is established outside the EU,
the EU authorized representative is:
Springer Nature Customer Service Center GmbH
Europaplatz 3, 69115 Heidelberg, Germany

Printed by Libri Plureos GmbH
in Hamburg, Germany